WORK RULES!

WORK RULES!

*Insights from Inside Google That Will
Transform How You Live and Lead*

LASZLO BOCK

TWELVE

NEW YORK   BOSTON

Twelve
Hachette Book Group
1290 Avenue of the Americas
New York, NY 10104

HachetteBookGroup.com

Printed in the United States of America

RRD-C

First Edition: April 2015
10 9 8 7 6 5 4 3

Twelve is an imprint of Grand Central Publishing.
The Twelve name and logo are trademarks of Hachette Book Group, Inc.

The Hachette Speakers Bureau provides a wide range of authors for speaking events. To find out more, go to www.hachettespeakersbureau .com or call (866) 376-6591.

The publisher is not responsible for websites (or their content) that are not owned by the publisher.

Library of Congress Cataloging-in-Publication Data

Bock, Laszlo.
    Work rules! : insights from inside Google that will transform how you live and lead / Laszlo Bock.
        pages cm
    Includes bibliographical references and index.
    ISBN 978-1-4555-5479-9 (hardback) — ISBN 978-1-4555-3484-5 (international trade paperback) — ISBN 978-1-4789-8087-2 (audio book) — ISBN 978-1-4789-8088-9 (audio download)   1. Leadership.   2. Corporate culture.   3. Management.   4. Google (Firm)—Management.   I. Title.
    HD57.7.B633 2015
    658.4'092—dc23
                    2014020843

For Annabelle, Emily, and Lila

may you always love what you do

Where's the work that'll set my hands, my soul free
—"WE TAKE CARE OF OUR OWN,"
BRUCE SPRINGSTEEN

Contents

WORK RULES!

Preface: A Guidance Counselor's Nightmare

Building the perfect Google resume, in retrospect

My first paycheck came in the summer of 1987, when I was fourteen years old. My best friend, Jason Corley, and I had been invited by our high school to enroll in a summer-school debate class the year before ninth grade. By the next year, we were teaching it. We earned $420 each.

Over the next twenty-eight years, I amassed a random walk resume that could best be described as a guidance counselor's nightmare: I worked in a deli, a restaurant, and a library. I tutored high school students in California and taught elementary school students English in Japan. I was first a lifeguard in real life at my college pool, and then I played one on TV, appearing on *Baywatch* as a 1960s lifeguard in a flashback and as that old acting standby, "Man walking across background." I helped start a nonprofit that supported troubled teens, and worked at a manufacturer that built construction products. I stumbled into consulting on executive pay, and with all the wisdom a twenty-four-year-old can muster, observed that human resources was a stagnant field and fled to get an MBA. Two years later I joined McKinsey & Company, the management consulting firm, where I focused as little on people issues as I could. During the dot-com boom years up until early 2000, I advised technology companies on how to grow sales, users, and organizations. And when the bubble popped, I advised technology companies on how to slash costs, run efficiently, and pivot into new businesses.

But by 2003 I was frustrated.

Frustrated because even the best-designed business plans fell apart when people didn't believe in them. Frustrated because leaders always spoke of putting people first, and then treated them like replaceable gears. (Low point on my first project: I asked my manager for career advice and he told me, "You guys are all like arrows in a quiver. Every one of you is the same.")

I'd held blue-collar jobs and white-collar jobs, been paid minimum wage and a six-figure salary, toiled alongside—and been managed by—people who didn't finish high school and people with PhDs from the poshest universities in the world. I had worked in an environment where our sole purpose was to change the world, and another where it was all about profits for the owner. It just didn't make sense to me that no matter where I turned, people weren't treated better in their jobs. You spend more time working than doing anything else in life.[1] It's not right that the experience of work, even at some of the best employers, should be so demotivating and dehumanizing.

I determined that I had two paths to choose from. The first was to treat my teams better, improve their output, and hope that over time others would follow my example. The second was to find a way to influence how entire companies treat people. I chose the latter path because I believed it would give me the greatest chance of affecting the most people, and decided to find a job in human resources (HR). My colleagues in consulting thought I was committing professional suicide, but I'd done my homework. At the time, there were more than five thousand people in McKinsey's database of alumni, but only a hundred of them were in human resources, virtually all working as consultants for other firms or recruiters. I reasoned that my training and background would make me stand out in the HR talent pool and help me come up with novel solutions. And maybe, just maybe, that would help me have a faster career trajectory than waiting twenty or thirty years to creep up the corporate ladder. I might get to a place where I could impact more people, faster.

I wanted to work at the places where I could learn as much as possible about HR, and Pepsi and General Electric were the best-regarded HR shops at the time. I cold-called eight HR executives from the two companies, but only one, Anne Abaya at GE, returned my call. Anne, a fluent Japanese speaker from Hawaii who somehow was always able to carve out a few minutes here and there to help people, found my background intriguing and introduced me to others at GE.

Six weeks later, I was hired. I was now the Vice President of Compensation and Benefits of the Commercial Equipment Financing division of the GE Capital division of the General Electric Company. I was thrilled to be there, though my friends took one look at my business card and thought I was nuts. My first boss, Michael Evans, gave me tremendous latitude to explore the company and helped me understand GE's approach to talent.

People mattered to Jack Welch, GE's chairman and CEO from 1981 to 2001. He spent more than 50 percent of his time on people issues,[2] and together with Bill Conaty, his chief human resources officer, built an acclaimed people management system by stringently ranking employees based on performance, choreographing job changes for top talent every twelve to eighteen months, and building a global training center in Crotonville, New York. Jack had handed over the reins to a new CEO, Jeff Immelt, two years before I joined, which allowed me to see what had been built and how it changed as Immelt's focus shifted to other areas.

Welch and Conaty had implemented a 20-70-10 performance ranking system, where GE employees were sorted into three groups: the top 20 percent, the middle 70 percent, and the bottom 10 percent. The top workers were lionized and rewarded with choice assignments, leadership training programs, and stock options. The bottom 10 percent were fired. Under Immelt, the forced distribution was softened and the crisp labels of "top 20 percent," "middle 70 percent," and "bottom 10 percent" were replaced with euphemisms: "top talent," "highly valued," and "needs improvement." Colleagues told me that the vaunted Session C process, a yearlong review of talent across the 300,000-person-strong company, had "lost its teeth" and "just wasn't the same without Jack's focus." [3]

I didn't have the benefit of having worked under both CEOs, but it dawned on me how deeply a CEO's persona and focus can shape an institution. Most CEOs are very good at many things, but they become CEOs for being superbly distinctive at one or two, which tend to be matched to a company's needs at that time. Even CEOs need to declare a major. Welch is best known for Six Sigma—a set of tools to improve quality and efficiency—and his focus on people. Immelt instead emphasized sales and marketing, most visibly through GE's branded "ecomagination" efforts to make and be perceived as a maker of greener products.

In 2006, after three years at GE, I was recruited to join Google as head of People Operations. I remember the recruiter, Martha Josephson, trying to convince me not to wear a suit to the interview. "No one wears suits," she assured me, "and they'll think you don't understand their culture if you show up in one." I took her advice, but was skeptical enough that I carried a necktie stuffed into my jacket pocket in case I needed it. Years later, I'd interview a candidate who had clearly purchased a beautiful pinstripe suit just for the interview, but who was still so exceptional that I could tell we'd hire him. I closed the interview with, "Brian, I have good news and bad news. The good news is that, while you still have more interviews to go, I can tell that you're going to get an offer. The bad news is that you're never going to get to wear that suit again."

When I joined, it was two years after Google's initial public offering: Revenues were growing 73 percent a year; Gmail had just launched with an unheard-of free gigabyte of storage (five hundred times more than prior webmail services—this was so crazy that people thought Gmail was an April Fool's joke)[4]; there were six thousand Google employees and the company wanted to double in size every year; and they had this wildly ambitious mission to organize the world's information—all of it!—and make it universally accessible and useful.

This mission for me was by far the most exciting part. I was born in 1972 in Communist Romania, a country ruled by the dictator Nicolae Ceausescu and permeated by secrecy, lies, and fear. It's hard to conceive of today, but Romania at the time was much like North Korea today. Friends and family members would disappear for criticizing the government. Members of the Communist Party had access to fine clothes, consumer goods, and fruit and vegetables from the West, while my parents didn't taste their first banana until they were in their thirties. Children were encouraged to spy on their parents. And the newspapers and radio disseminated little but lies about how great the government was and how evil and oppressive the United States was. My family fled Romania seeking freedom, the right to go where they wanted, say and think what they wanted, associate with whomever they wanted.

The idea of joining a company founded with a goal of making information available to everyone was thrilling, because the state of freedom is predicated on free expression, which in turn relies on access to information and truth. I'd lived and worked in all kinds of environments and seen lots of examples of what didn't work. If this place is for real, I thought, this is going to be the best job in the world.

Since I joined, Google has grown from six thousand employees to almost sixty thousand, with seventy-plus offices across more than forty countries. *Fortune* has named Google the "Best Company to Work For" an unprecedented five times in the United States, as well as numerous times in countries as diverse as Argentina, Australia, Brazil, Canada, France, India, Ireland, Italy, Japan, Korea, the Netherlands, Poland, Russia, Switzerland, and the UK. Google is the most sought-after place to work on the planet according to LinkedIn,[5] and we receive more than two million applications every year, representing individuals from every background and part of the world. Of these, Google hires only several thousand per year,[6] making Google twenty-five times more selective than Harvard,[7] Yale,[8] or Princeton.[9]

Far from being professional suicide, my time at Google has been a white-water ride of experimentation and creation. Sometimes exhausting, sometimes frustrating, but always surging forward to create an environment of purpose, freedom, and creativity. This book is the story of how we think about our people, what we've learned over the past fifteen years, and what you can do to put people first and transform how you live and lead.

Why Google's Rules Will Work for You

The surprising (and surprisingly successful) places that work just as we do

A billion hours ago, modern Homo sapiens emerged.
A billion minutes ago, Christianity began.
A billion seconds ago, the IBM personal computer was released.
A billion Google searches ago...was this morning.

—HAL VARIAN, GOOGLE'S CHIEF
ECONOMIST, DECEMBER 20, 2013

Google turned sixteen years old in 2014, but became part of the fabric of our lives long before. We don't search for something on the Internet, we "Google it." More than a hundred hours of video are uploaded to YouTube every minute. Most mobile phones and tablets rely on Google's free, open-source[i] operating system, Android, which didn't exist in the market before 2007. More than fifty billion apps have been downloaded from the Google Play store. Chrome, launched as a safer, faster, and open-source Web browser in 2008, has over 750 million active users and has grown into an operating system powering "Chromebook" laptops.[10]

---

[i] "Open source" means the software is freely available and can be modified. For example, Amazon's e-book reader, Kindle, runs on a modified version of the Android operating system.

And Google is just beginning to explore what is possible, from self-driving cars to Project Loon, which aims to provide Internet access by balloon to the hardest-to-reach parts of the globe. From wearable computing products like Google Glass, which blends the Web and the world in a tiny lens that sits above your right eye (we're working on a version for lefties), to the Google Smart Contact Lens, a contact lens that doubles as a blood glucose monitor for people with diabetes.

Each year, tens of thousands of visitors come to our campuses around the world. They include social and business entrepreneurs, high school and college students, CEOs and celebrities, heads of state and kings and queens. And of course, our friends and families, who are always happy to stop by for a free lunch. They all ask about how we run this place, about how Google works. What is the culture all about? How do you actually get any work done with all the distractions? Where does the innovation come from? Do people really get 20 percent of their time to do whatever they want?

Even our employees, "Googlers" as they call themselves, sometimes wonder why we do things a certain way. Why do we spend so much time on recruiting? Why do we offer some perks and not others?

Work Rules! is my attempt to answer those questions.

Inside Google we don't have a lot of rule books and policy manuals, so this isn't the official corporate line. Instead, it's my interpretation of why and how Google works, viewed through the lens of what I believe to be true—and what the latest research in behavioral economics and psychology has revealed—about human nature. As the SVP of People Operations, it continues to be a privilege and delight to play a role, along with a cast of literally thousands of Googlers, in shaping how Googlers live and lead.

The first time Google was named the Best Company to Work For in the United States was a year after I joined (not thanks to me— but my timing was good). The sponsors of the award, *Fortune* magazine and the Great Place to Work Institute, invited me to sit on a panel with Jack DePeters, SVP of store operations at Wegmans, an eighty-four-store grocery chain in the northeastern United States that has earned a seventeen-year run on *Fortune*'s list of best companies to work for, taking the top spot in 2005 and showing up in the top five almost every year since.[11]

The point of having us both on stage was to showcase our distinctive management philosophies, to show that there was more than one path to becoming a superb employer. Wegmans is a privately held regional retailer that operates in an industry with an average 1 percent profit margin, and its largely local workforce has for the most part a high school education. They've been around since 1916 and have been family-run the whole time. Google at the time was a nine-year-old publicly held global technology company with a roughly 30 percent profit margin; its recruits, drawn from all over the world, collected PhDs like trading cards. The two companies could not have been more different.

I was stunned to learn that our companies had far more in common than not.

Jack explained that Wegmans adheres to virtually the same principles as Google: "Our CEO, Danny Wegman, says that 'leading with your heart can make a successful business.' Our employees are empowered around this vision to give their best and let no customer leave unhappy. And we use it to always make our decisions to do the right thing with our people, regardless of cost."

Wegmans gives employees full discretion to take care of customers, awarded $5.1 million in scholarships to employees in 2013,[12] and even encouraged an employee to start her own in-store bakery simply because her homemade cookies were so good.

Over time I learned that Wegmans and Google weren't alone in their approach. The Brandix Group is a Sri Lankan clothing manufacturer, with more than forty plants in Sri Lanka and substantial operations in India and Bangladesh. Ishan Dantanarayana, their chief people officer, told me that their goal is "inspiring a large female workforce" by telling employees to "come as you are and harness your full potential." In addition to making their CEO and board accessible to all employees, they provide pregnant women with supplemental food and medicine; offer a diploma program that allows employees to learn as they work and even trains them to be entrepreneurs and start their own businesses; appoint worker councils in all plants to help every employee influence the business; offer scholarships for children of employees; and more. They also give back to the community, for example through their Water & Women program, which builds wells in employees' villages. "This elevates the stature of our employees in the community, and they are then privy to clean water, which is scarce."

All these efforts have made them Sri Lanka's second-largest exporter and the recipient of numerous awards for their employment conditions, community involvement, and environmental practices. Ishan elaborated how this happens: "When employees trust the leadership, they become brand ambassadors and in turn cause progressive change in their families, society, and environment. The return on investment to business is automatic, with greater productivity, business growth, and inspired customers."

Contrast Brandix's approach with the collapse of the Rana Plaza building in Bangladesh on April 24, 2013. Five apparel manufacturers, a bank, and several shops filled the eight-floor building. The day before, Rana Plaza was evacuated as cracks appeared in the walls. The next day, the bank and shops told their employees to stay away. The apparel companies ordered their workers back in. More than 1,100 people lost their lives, including children who were in a company nursery in the building.[13]

Closer to home, the 1999 film *Office Space*, which deadpanned the meaningless rituals and bureaucracy of a fictional technology company, became a cult hit because it was instantly recognizable.

In the movie, programmer Peter Gibbons describes his job hypnotherapist:

> *Peter:* So I was sitting in my cubicle today, and I realized, ever since I started working, every single day of my life has been worse than the day before it. So that means that every single day that you see me, that's on the worst day of my life.
>
> *Dr. Swanson:* What about today? Is today the worst day of your life?
>
> *Peter:* Yeah.
>
> *Dr. Swanson:* Wow, that's messed up.[14]

I thought of these vastly different examples when a reporter from CNN International called for an article about the future of work. She argued that the model exemplified by places like Google—what I'll call a "high-freedom" approach where employees are given great latitude—was the way of the future. Top-down, hierarchical, command-and-control models of management— "low-freedom" environments—would soon fade away.

Someday, perhaps. But soon? I wasn't so sure. Command-oriented, low-freedom management is common because it's profitable, it requires less effort, and most managers are terrified of the alternative. It's easy to run a team that does what they are told. But to have to explain to them why they're doing something? And then debate whether it's the right thing to do? What if they disagree with me? What if my team doesn't want to do what I tell them to? And won't I look like an idiot if I'm wrong? It's faster and more efficient to just tell the team what to do and then make sure they deliver. Right?

Wrong. The most talented people on the planet are increasingly physically mobile, increasingly connected through technology, and—importantly—increasingly discoverable by employers. This

o be in high-freedom companies, and talent will
panies. And leaders who build the right kind of
be magnets for the most talented people on the

uilding such a place, because the power dynamic
nagement pulls against freedom. Employees are
dependent on their managers and want to please them. A focus on
pleasing your manager, however, means it can be perilous to have
a frank discussion with her. And if you don't please her, you can
become fearful or resentful. At the same time, she's accountable for
you delivering certain results. Nobody produces their best work
entangled in this Gordian knot of spoken and unspoken agendas
and emotions.

Google's approach is to cleave the knot. We deliberately take
power and authority over employees away from managers. Here is a
sample of the decisions managers at Google cannot make unilaterally:

- Whom to hire
- Whom to fire
- How someone's performance is rated
- How much of a salary increase, bonus, or stock grant to give
  someone
- Who is selected to win an award for great management
- Whom to promote
- When code is of sufficient quality to be incorporated into
  our software code base
- The final design of a product and when to launch it

Each of these decisions is instead made either by a group of
peers, a committee, or a dedicated, independent team. Many newly
hired managers hate this! Even once they get their heads around
the way hiring works, promotion time comes around and they are
dumbfounded that they can't unilaterally promote those whom they

believe to be their best people. The problem is that you and I might define our "best people" differently. Or it might be possible that your worst person is better than my best person, in which case you should promote everyone and I should promote no one. If you're solving for what is most fair across the entire organization, which in turn helps employees have greater trust in the company and makes rewards more meaningful, managers must give up this power and allow outcomes to be calibrated across groups.

What's a manager to do without these traditional sticks and carrots? The only thing that's left. "Managers serve the team," according to our executive chairman, Eric Schmidt. Like any place, we of course have exceptions and failures, but the default leadership style at Google is one where a manager focuses not on punishments or rewards but on clearing roadblocks and inspiring her team. One of our lawyers described his manager, Terri Chen, this way: "You know that killer line from *As Good As It Gets* where Jack Nicholson says to Helen Hunt: 'You make me want to be a better man'? That is how I feel about Terri as a manager. She makes me want to—and helps me try to be—a better Googler and trademark lawyer and person!" The irony is that the best way to arrive at the beating heart of great management is to strip away all the tools on which managers most rely.

The good news is that any team can be built around the principles that Google has used. Even in the garment industry, MIT's Richard Locke found that this kind of approach works.[15] He compared two Nike T-shirt factories in Mexico. Plant A gave workers more freedom, asking them to help set production targets, organize themselves into teams, and decide how work would be broken up, and granting them authority to stop production when they saw problems. Plant B tightly controlled the shop floor, requiring workers to stick to their assigned tasks and adding strict rules about when and how work happened. Locke found that workers at Plant A were almost twice as productive (150 T-shirts per day vs. 80), earned higher wages, and had 40 percent lower costs per T-shirt ($0.11/shirt vs. $0.18/shirt).

Dr. Kamal Birdi of the University of Sheffield and six other researchers studied the productivity of 308 companies across twenty-two years and came to a similar conclusion. These companies had all launched traditional operations programs like "total quality management" and "just-in-time inventory control." Birdi found that these programs sometimes improved productivity in one company or another, but "we found no overall performance effect" when the companies were looked at in aggregate. In other words, there was no evidence suggesting that any of these operations initiatives would reliably and consistently improve performance.

So what did? Performance improved only when companies implemented programs to empower employees (for example, by taking decision-making authority away from managers and giving it to individuals or teams), provided learning opportunities that were outside what people needed to do their jobs, increased their reliance on teamwork (by giving teams more autonomy and allowing them to self-organize), or a combination of these. These factors "accounted for a 9% increase in value added per employee in our study." In short, only when companies took steps to give their people more freedom did performance improve.[16]

That's not to say that Google's approach is perfect or that we don't make our fair share of mistakes. We've taken some bruises along the way, as you'll see in chapter 13. I expect my examples and arguments will be greeted with a healthy helping of skepticism in some quarters. All I can say in my defense is that this is really how it works at Google, and this is really why we run the company this way. And a kindred approach works for Brandix, Wegmans, and dozens of other organizations and teams, both large and small.

I once gave a talk in Chicago to a group of local chief human resources officers (CHROs) about Google's culture. After the presentation, one CHRO stood up and sneered, "This is all well and good for Google. You have huge profit margins and can afford to treat your people so well. We can't all do that."

I was going to explain that most of what we did cost us little to nothing. And that even in a time of flat wages you can still make work better, make people happier. Indeed, it's when the economy is at its worst that treating people well matters most.

Before I could muster a response, another CHRO argued back, "What do you mean? Freedom is free. Any of us can do this."

He was right.

All it takes is a belief that people are fundamentally good— and enough courage to treat your people like owners instead of machines. Machines do their jobs; owners do whatever is needed to make their companies and teams successful.

People spend most of their lives at work, but work is a grinding experience for most—a means to an end. It doesn't have to be.

We don't have all the answers, but we have made some fascinating discoveries about how best to find, grow, and keep people in an environment of freedom, creativity, and play.

The secrets of Google's people success can be replicated in organizations large and small, by individuals and CEOs. Not every company will be able to duplicate perks like free meals, but everyone can duplicate what makes Google great.

Becoming a Founder

Just as Larry and Sergey laid the foundation for
how Google treats its people, you can lay the
foundation for how your team works and lives

Every great tale starts with an origin story.

The infants Romulus and Remus, abandoned beside the
Tiber River, are nursed by a she-wolf, fed by a woodpecker,
and then raised by kindly shepherds. As a young man, Romulus
goes on to found the city of Rome.

Baby Kal-El rockets to earth as his home planet Krypton
explodes behind him, landing in Smallville, Kansas, to be raised by
the kindly Martha and Jonathan Kent. Moving to Metropolis, he
takes on the mantle of Superman.

Thomas Alva Edison opens a lab in Menlo Park, New Jersey,
in 1876. He brings together an American mathematician, an English machinist, a German glassblower, and a Swiss clockmaker who
develop an incandescent lightbulb that burns for more than thirteen hours,[17] laying the foundation for the Edison General Electric
Company.

Oprah Winfrey, born of an impoverished teenage mother,
abused as a child, and shuttled from home to home, goes on to
become an honors student, the youngest and first black news anchor
at WLAC-TV in Nashville, and one of the most successful communicators and inspirational businesspeople in the world.[18]

Vastly different tales, yet all teasingly similar. The mythologist Joseph Campbell argued that there are just a few archetypal stories that underpin most myths around the world. We are called to an adventure, face a series of trials, become wiser, and then find some manner of mastery or peace. We humans live through narrative, viewing history through a lens of stories that we tell ourselves. No wonder that we find common threads in the tapestries of one another's lives.

Google has an origin story too. Most think it began when Larry Page and Sergey Brin, Google's founders, met during a campus tour for new students at Stanford University. But it starts much earlier than that.

Larry's views were shaped by his family history: "My grandfather was an autoworker, and I have a weapon he manufactured to protect himself from the company that he would carry to work. It's a big iron pipe with a hunk of lead on the head."[19] He explained, "The workers made them during the sit-down strikes to protect themselves."[20]

Sergey's family had defected from the Soviet Union in 1979, seeking freedom and a respite from the anti-Semitism of the Communist regime. "My rebelliousness, I think, came out of being born in Moscow," explained Sergey. "I'd say this is something that followed me into adulthood."[21]

Larry's and Sergey's ideas about how work could be were also informed by their early experiences at school. As Sergey has commented: "I do think I benefited from the Montessori education, which in some ways gives the students a lot more freedoms to do things at their own pace." Marissa Mayer, at the time a Google vice president of product management and now CEO of Yahoo, told Steven Levy in his book *In the Plex*: "You can't understand Google... unless you know that both Larry and Sergey were Montessori kids."[22] This teaching environment is tailored to a child's learning needs and personality, and children are encouraged to question everything, act of their own volition, and create.

In March 1995, a twenty-two-year-old Larry Page was visiting Stanford University in Palo Alto, California. He was finishing his undergraduate degree at the University of Michigan and considering entering Stanford's PhD program in computer science. Sergey, twenty-one years old, had graduated from the University of Maryland two years earlier[ii] and was already enrolled in the PhD program. He was volunteering as a tour guide for prospective students. And of course, Larry was assigned to Sergey's tour group.[23]

They quickly developed a friendly banter, and a few months later Larry showed up as a new student. Larry was fascinated with the World Wide Web, and particularly the way Web pages connected to one another.

The Web in 1996 was a chaotic mess. In simplest terms, search engines wanted to show the most relevant, useful Web pages, but ranked them mainly by comparing the text on a Web page to the search query that was typed. That left a loophole. The owner of a Web page could boost his rankings on search engines with tricks like hiding popular search terms in invisible text on the page. If you wanted people to come to your pet food site, you could write "pet food" in blue text on a blue background a hundred times, and your search ranking would improve. Another trick was to repeat words again and again in the source code that generated your page but was invisible to a human reader.

Larry reasoned that an important signal was being overlooked: what users thought of the Web page. The most useful Web pages would have lots of links from other sites, because people would link only to the most useful pages. That signal would prove to be far more powerful than the words written on the page itself.

[ii] Sergey left high school a year early and finished college in three years.

But creating a program that could identify every link on the Web and then tabulate the strength of every relationship across all websites at the same time was an inhumanly complex problem. Fortunately, Sergey found the problem equally captivating. They created BackRub, a reference to the backlinks reaching back from the site you saw to the site you had just been on. In August 1998, Andy Bechtolsheim, one of the cofounders of Sun Microsystems, famously wrote a $100,000 check to "Google, Inc." before the company was even incorporated. Less well known is that they moments later received a second $100,000 check from Stanford professor David Cheriton, on whose porch they had met Andy.[24]

Reluctant to leave Stanford to start a company, Larry and Sergey tried to sell Google but were unable to. They offered it to AltaVista for $1 million. No luck. They turned to Excite and at the urging of Vinod Khosla, a partner at venture capital firm Kleiner Perkins Caufield & Byers, lowered the price to $750,000. Excite passed.[iii]

This was before Google's first advertising system, AdWords, was launched in 2000, before Google Groups (2001), Images (2001), Books (2003), Gmail (2004), Apps (spreadsheets and documents for businesses, 2006), Street View (2007), and dozens of other products we use every day. It was before Google Search was available in over 150 languages, and before we opened our first international office in Tokyo (2001). And way before your Android phone could buzz you in advance if your flight was delayed, or you could say to the Google Glass on your eyeglass frame, "Okay, Glass, take a picture and send it to Chris," and know Chris will get to see through your eyes.

[iii] One of the great lessons of Google's history is that to succeed you need to have a brilliant idea, great timing, exceptional people…and luck. Though it didn't feel like it at the time, failing to sell the company was a tremendously fortunate break, as were the chance meeting between Larry and Sergey on a campus tour and dozens of other events. It would be easy to claim that our success was due to being somehow smarter or working harder, but that's just not true. Smarts and hard work are necessary but not sufficient conditions for success. We got lucky too. Kinda gives the "I'm Feeling Lucky" button on our home page a whole new meaning.

Larry and Sergey had ambitions beyond developing a great search engine. They started out knowing how they wanted people to be treated. Quixotic as it sounds, they both wanted to create a company where work was meaningful, employees felt free to pursue their passions, and people and their families were cared for. "When you're a grad student," Larry observed, "you can work on whatever you want. And the projects that were really good got a lot of people really wanting to work on them. We've taken that learning to Google, and it's been really, really helpful. If you're changing the world, you're working on important things. You're excited to get up in the morning. You want to be working on meaningful, impactful projects, and that's the thing there is really a shortage of in the world. I think at Google we still have that."

Many of the most meaningful, beloved, and effective people practices at Google sprouted from seeds planted by Larry and Sergey. Our weekly all-employee meetings started when "all" of us amounted to just a handful of people, and continue to this day even though we're now the size of a respectable city. Larry and Sergey always insisted that hiring decisions be made by groups rather than a single manager. Employees calling meetings simply to share what they were working on turned into the hundreds of Tech Talks we host each month. The founders' early generosity led to an almost unprecedented sharing of ownership in the company: Google is one of the few companies of our size to grant stock to all employees. Our efforts to draw more women into computer science started before we had thirty employees, and at Sergey's direct request. Our policy of welcoming dogs at work originated with our first ten people. (As did our position on cats, which is enshrined in our code of conduct: "We like cats, but we're a dog company, so as a general rule we feel cats visiting our offices would be fairly stressed out."[25]) And of course, our tradition of free meals started with free cereal and an enormous bowl of M&M's.

When Google went public on August 19, 2004, Sergey included

a letter in our prospectus for investors, describing how the founders felt about their 1,907 employees. The italics are his:

> *Our employees, who have named themselves Googlers, are every-thing. Google is organized around the ability to attract and lever-age the talent of exceptional technologists and business people. We have been lucky to recruit many creative, principled and hard working stars. We hope to recruit many more in the future. We will reward and treat them well.*
>
> We provide many unusual benefits for our employ-ees, including meals free of charge, doctors and washing machines. We are careful to consider the long-term advan-tages to the company of these benefits. Expect us to add benefits rather than pare them down over time. We believe it is easy to be penny wise and pound foolish with respect to benefits that can save employees considerable time and improve their health and productivity.
>
> The significant employee ownership of Google has made us what we are today. Because of our employee talent, Google is doing exciting work in nearly every area of com-puter science. We are in a very competitive industry where the quality of our product is paramount. Talented people are attracted to Google because we empower them to change the world; Google has large computational resources and distribution that enables individuals to make a difference. Our main benefit is a workplace with important projects, where employees can contribute and grow. We are focused on providing an environment where talented, hard working people are rewarded for their contributions to Google and for making the world a better place.

Google was fortunate that our founders had such strong beliefs about the kind of company they wanted to create.

But Larry and Sergey weren't the first.

Henry Ford is best known for his sweeping adoption of the assembly line. It's less well known that his philosophy of recognizing and rewarding work was remarkably progressive for the time:

> The kind of workman who gives the business the best that is in him is the best kind of workman a business can have. And he cannot be expected to do this indefinitely without proper recognition.... [I]f a man feels that his day's work is not only supplying his basic need, but is also giving him a margin of comfort, and enabling him to give his boys and girls their opportunity and his wife some pleasure in life, then his job looks good to him and he is free to give it of his best. This is a good thing for him and a good thing for the business. The man who does not get a certain satisfaction out of his day's work is losing the best part of his pay.[26]

This is entirely consistent with Google's view, though Henry Ford wrote these words more than ninety years ago, in 1922. And he acted on them as well, doubling the wages of his factory workers in 1914 to $5 per day.

Even earlier, in 1903, Milton S. Hershey not only laid the foundation for what would become the Hershey Company but also for the town of Hershey, Pennsylvania. The United States had over 2,500 company towns in the nineteenth and early twentieth centuries, housing 3 percent of the population at their peak.[27] But unlike in most company towns, Hershey "avoided building a faceless company town with row houses. He wanted a 'real home town' with tree-lined streets, single- and two-family brick houses, and manicured lawns."

> With Milton Hershey's success came a profound sense of moral responsibility and benevolence. His ambitions were not limited to producing chocolate. Hershey envisioned a

complete new community around his factory. He built a model town for his employees that included comfortable homes, an inexpensive public transportation system, a quality public school system and extensive recreational and cultural opportunities.[28]

Which is not to say that all of Ford's and Hershey's views were palatable. Some were abhorrent. Ford was widely criticized for publishing anti-Semitic works and later apologized.[29] Hershey too allowed racist commentary to be published under his leadership in the Hershey town newspaper.[30] But it's also clear that—at least for a subset of people—both of these founders saw value in considering workers as something more than manufacturing inputs.

A more recent, and less morally ambiguous, example is Mervin J. Kelly, who joined Bell Labs in 1925 and served as president from 1951 to 1959.[31] During his tenure, Bell Labs invented lasers and solar cells, laid the first transatlantic phone cable, developed crucial technologies that made possible the rise of the microchip, and created the foundation for information theory through its work on binary code systems. This built on Bell Labs' earlier work, which included the invention of the transistor in 1947.

Upon becoming president, Kelly took an unorthodox approach to management. First, he upended the physical design of their Murray Hill, New Jersey, labs. Rather than a traditional layout with each floor segregated into sections for each specialized area of research, Kelly insisted on a floor plan that forced interaction across departments: Offices were along long corridors spanning the entire floor, so that walking down the hall all but guaranteed that colleagues would stumble over each other and be drawn into one another's work. Second, Kelly built Franken-teams, combining "thinkers and doers" as well as disparate experts on single teams. Author Jon Gertner described one such team in his history of Bell Labs, *The Idea Factory*:[32] "Purposefully mixed together on the transistor project were physicists, metallurgists and electrical engineers; side by side were specialists in theory, experimentation and manufacturing."

Third, Kelly gave people freedom. Gertner continues:

Mr. Kelly believed that freedom was crucial, especially in research. Some of his scientists had so much autonomy that he was mostly unaware of their progress until years after he authorized their work. When he set up the team of researchers to work on what became the transistor, for instance, more than two years passed before the invention occurred. Afterward, when he set up another team to handle the invention's mass manufacture, he dropped the assignment into the lap of an engineer and instructed him to come up with a plan. He told the engineer he was going to Europe in the meantime.

Kelly's case is particularly fascinating because he wasn't the founder of Bell Labs, or even a rapidly rising star. On the contrary, he quit twice because he felt his projects weren't adequately funded (and in both cases was lured back with the promise of more funding). He was mercurial and had a nasty temper. An early manager, H. D. Arnold, "kept him for a long time at a low administrative level because he distrusted his judgment."[33] As a result, his career moved slowly. He worked as a physicist for twelve years before becoming director of Vacuum Tube Development, and it was another six years before he was made director of research. He was made president of Bell Labs twenty-six years after joining Bell.

What I love about his story is that Kelly acted like a founder. Like an owner. He didn't care only about the output of Bell Labs; he cared about the kind of place it was. He wanted brilliance to work free from the scrutinizing eye of management, while being constantly jostled by the elbows of the geniuses down the hall. It wasn't his job to care about building design and foot traffic patterns, but in doing so he became the spiritual founder of one of the most innovative organizations in history.[iv]

Turning back to Google, Larry and Sergey deliberately left space for others to act as founders. People with vision were given the opportunity to create their own Google. For years, the troika of Susan Wojcicki, Salar Kamangar, and Marissa Mayer were referred to as the "mini-founders," critical early Googlers who would go on to build and lead our advertising, YouTube, and search efforts, in partnership with brilliant computer scientists such as Sridhar Ramaswamy, Eric Veach, Amit Singhal, and Udi Manber. Craig Nevill-Manning, a gifted engineer, opened up our New York office because he preferred the big city to the suburbs of Silicon Valley. Omid Kordestani, recruited out of his job as the top sales exec at Netscape to build and lead the sales team for Google, is often referred to by Larry, Sergey, and Eric Schmidt as the "business founder" of Google. Fast-forward more than a decade, and Googlers are still acting like owners: Craig Cornelius and Richard Treitel decided to make a Cherokee language interface for Google, helping in a small way to preserve an endangered language.[34] Ujjwal Singh, Steve Crossan, and AbdelKarim Mardini partnered with engineers from Twitter following the Egyptian government's shutdown of the Internet in early 2011 to create Speak2Tweet, a product that takes messages from a voice mailbox and transcribes them into Tweets broadcast around the world.[35] This gave Egyptians a way to com-

---

[iv] It wasn't just the boys who were building a different kind of workplace. In Paris, fashion designer and entrepreneur Madeleine Vionnet went to work as an apprentice seamstress at the age of eleven. In 1912, at the age of thirty-six, she founded her eponymous fashion house and in the next decade introduced the bias cut (*coup en bias*), replacing corsets with slim, body-hugging fabrics. Even in the midst of the Great Depression, her employees received "free medical and dental care, maternity leave and babysitting services, and paid holidays, too," according to Northwestern University professor Deborah Cohen. [Sources: http://www.vionnet.com/madeleine-vionnet; http://www.theatlantic.com/magazine/archive/2014/05/the-way-we-look-now/359803/.]

municate en masse with the world and, by dialing into the voice mailbox, to listen to one another.

You are a founder

Building an exceptional team or institution starts with a founder. But being a founder doesn't mean starting a new company. It is within anyone's grasp to be the founder and culture-creator of their own team, whether you are the first employee or joining a company that has existed for decades.

At Google, we don't believe we've stumbled on the only model for people success. We certainly don't have all the answers. And we absolutely screw up much more often than we'd hope to. But we have been able to prove that many of Larry and Sergey's original instincts were right, to debunk some management lore, and to discover some shocking things along the way. Our aspiration is that in some small way, sharing the lessons we've learned will improve how people experience work everywhere.

The Russian novelist Leo Tolstoy wrote, "All happy families resemble one another."[v] All successful organizations resemble one another as well. They possess a shared sense not just of what they produce, but of who they are and want to be. In their vision (and perhaps hubris), they've thought through not just their origin, but also their destiny.

One of my hopes in writing this book is that anyone reading it starts thinking of themselves as a founder. Maybe not of an entire company, but the founder of a team, a family, a culture. The fundamental lesson from Google's experience is that you must first choose whether you want to be a founder or an employee. It's not a question of literal ownership. It's a question of attitude.

---

[v] Leo Tolstoy, *Anna Karenina*. He concluded morosely: "Every unhappy family is unhappy in its own way."

In Larry's words: "I think about how far we've come as companies from those days, where workers had to protect themselves from the company. My job as a leader is to make sure everybody in the company has great opportunities, and that they feel they're having a meaningful impact and are contributing to the good of society. As a world, we're doing a better job of that. My goal is for Google to lead, not follow."[36]

That's how a founder thinks.

Student or senior executive, being part of an environment where you and those around you will thrive starts with your taking responsibility for that environment. This is true whether or not it's in your job description and whether or not it's even permitted.

And the greatest founders create room for other founders to build alongside them.

One day your team will have an origin story, a founding myth, just like Rome or Oprah or Google. Think about what you want it to be, about what you want to stand for. Think about what stories people will tell about you, your work, your team. Today you have the opportunity to become the architect of that story. To choose whether you want to be a founder or an employee.

I know which I'd choose.

....

WORK RULES... FOR BECOMING A FOUNDER

☐ Choose to think of yourself as a founder.
☐ Now act like one.

....

2

....

"Culture Eats Strategy for Breakfast"

If you give people freedom, they will amaze you

I receive a lot of oddball mail at work, usually from people who want to work at Google. I've received T-shirts with resumes silk-screened on them, puzzles, and even sneakers (from someone wanting to "get their foot in the door"...get it?). I post the more colorful ones on my wall, including one letter that included the phrase "culture eats strategy for breakfast." I'd never heard the phrase but thought it silly enough to keep as an example of management gibberish.

If you do a Google image search for "Google culture" you'll get something like the screenshot on the next page.

These pictures encapsulate how first-time visitors perceive Google's culture. The colorful slides and beanbags, the free gourmet food, the crazy offices (yes, that's someone riding a bike through the office), and happy people working together and having a great time all suggest that this place is about work-as-play. There's an element of truth to that, but Google's culture has much deeper roots. Ed Schein, now retired from the MIT School of Management, taught that a group's culture can be studied in three ways: by looking at its "artifacts," such as physical space and behaviors; by surveying the beliefs and values espoused by group members; or

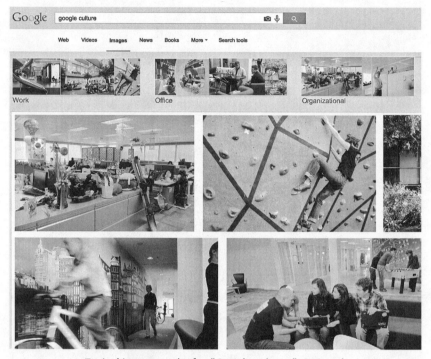

Typical image results for "Google culture." © Google, Inc.

by digging deeper into the underlying assumptions behind those values.[37] It's natural to look at Google and focus on the physical spaces: the nap pods for catching a quick snooze or the slides connecting floors. As Adam Grant of the University of Pennsylvania's Wharton School and its youngest-ever tenured professor told me: "People interpret strong cultures based on the artifacts, because they're the most visible, but the values and assumptions underneath matter much more."

Adam's right.

"Fun" is in fact the most common word Googlers use to describe our culture.[38] (Normally I'd be skeptical of employees telling the company how much they love it, but these surveys are anonymous—if anything, the incentive is to make things sound worse!) Early on, we decided that "you can be serious without a suit," and enshrined that notion in the "10 Things We Know to Be True," a list of ten beliefs that guide how we run our business.[vi]

We even play with our brand, something that most companies hold sacrosanct, swapping our regular logo for Google Doodles on our website. The first one, on August 30, 1998, was a tongue-in-cheek out-of-office notice for Larry and Sergey:

The Burning Man Google Doodle. © Google, Inc.

They had gone to Burning Man, an annual festival of art, community, and self-reliance in the Nevada desert. The figure in the middle represented the Burning Man himself.

On June 9, 2011, we commemorated Les Paul, one of the pioneers of the solid-body electric guitar, with an interactive Doodle. If you brushed the guitar strings with your mouse or finger, you made music. You could even hit the red button to record and share your song. By some estimates, visitors to our site spent over 5.3 million hours making music that day.[39]

[vi] The 10 Things are: Focus on the user and all else will follow. It's best to do one thing really, really well. Fast is better than slow. Democracy on the Web works. You don't need to be at your desk to need an answer. You can make money without doing evil. There's always more information out there. The need for information crosses all borders. You can be serious without a suit. Great just isn't good enough.

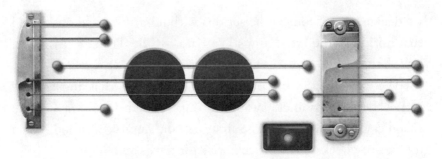

The Les Paul Google Doodle. © Google, Inc.

We celebrate April Fool's Day each year. On April 1, 2013, we announced that YouTube was in reality an eight-year contest to find the best video ever created, and we were finally ready to announce a winner. My favorite prank was Google Translate for Animals, an Android application in the UK that would translate animal sounds into English (Android is our operating system for mobile devices). It was not to be confused with some of the languages that Google Translate actually did convert into, such as Swedish Chef (from *The Muppet Show*—"Bork, bork, bork!") or Pirate ("Arrr!"). On April 1, 2012, anyone searching for music in the Google Play online store would see an Androidified Kanye West pop up and say, "Did you mean: Beyoncé."

We also have fun with our ongoing products. Each year, we launch a Santa Tracker so children can track Santa Claus as he travels the world. And watch what happens if you type "do a barrel roll" next time you're using Google.com or in a Chrome browser.

All this fun might sound too frivolous to take seriously, but fun is an important part of Google, creating an opportunity for unguarded exploration and discovery. Yet fun is an outcome of who we are, rather than the defining characteristic. It doesn't explain how Google works or why we choose to operate the way we do. To understand that, you have to explore the three defining aspects of our culture: mission, transparency, and voice.

A mission that matters

Google's mission is the first cornerstone of our culture. Our mission is "to organize the world's information and make it universally accessible and useful."[40] How does our mission compare to those of other companies? Here are a few excerpts from other companies in 2013 (emphasis added):

> IBM: "We strive to lead in the invention, development and manufacture of the industry's most advanced information technologies, including computer systems, software, storage systems and microelectronics. *We translate these advanced technologies into value for our customers* through our professional solutions, services and consulting businesses worldwide."[41]
>
> McDonald's: "McDonald's brand mission is to be our customers' favorite place and way to eat and drink. Our worldwide operations are aligned around a global strategy called the Plan to Win, which center on an exceptional customer experience—People, Products, Place, Price and Promotion. *We are committed to continuously improving our operations and enhancing our customers' experience*."[42]
>
> Procter & Gamble: "We will provide branded products and services of superior quality and value that improve the lives of the world's consumers, now and for generations to come. As a result, *consumers will reward us with leadership sales, profit and value creation*, allowing our people, our shareholders and the communities in which we live and work to prosper."[43]

These are all perfectly reasonable, responsible missions.

But two things are immediately obvious from reading these. First, I owe you an apology for making you slog through corporate mission statements, perhaps the worst form of literature known to man. Second, Google's mission is distinctive both in its simplicity and in what it doesn't talk about. There's no mention of profit or market. No mention of customers, shareholders, or users. No mention of why this is our mission or to what end we pursue these goals. Instead, it's taken to be self-evident that organizing information and making it accessible and useful is a good thing.

This kind of mission gives individuals' work meaning, because it is a moral rather than a business goal. The most powerful movements in history have had moral motivations, whether they were quests for independence or equal rights. And while I don't want to push this notion too far, it's fair to say that there's a reason that revolutions tend to be about ideas and not profits or market share.

Crucially, we can never achieve our mission, as there will always be more information to organize and more ways to make it useful. This creates motivation to constantly innovate and push into new areas. A mission that is about being "the market leader," once accomplished, offers little more inspiration. The broad scope of our mission allows Google to move forward by steering with a compass rather than a speedometer. While there are always disagreements—and we'll get to a few of those in chapter 13—the underlying shared belief in this mission unites most Googlers. It provided a touchstone for keeping the culture strong, even as we grew from dozens of people to tens of thousands.

One example of our mission pushing us into unexpected areas is Google Street View, which was introduced in 2007.[44] Its simple but mind-bogglingly expansive purpose was to create a historical record by documenting what the entire world looked like from street level. It built upon the success of Google Maps, which in turn was built on the foundation laid by John Hanke and Brian McClendon, who started a company called Keyhole in 2001 that Google acquired three years later (and who are both still vice presidents at Google).

After a few years of looking at overhead maps, Larry had asked why we couldn't also capture imagery the way people really saw it—from the ground. That was information too, and we'd be able to see how communities grow and change over time. And perhaps something interesting would come of it.

Something did.

The Arc de Triomphe!

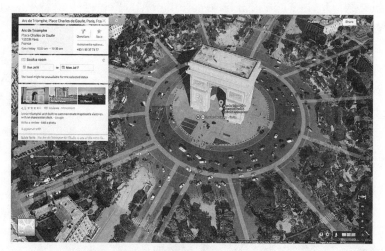

A sky-high view of the Arc de Triomphe in Paris, France, from Google Maps. © Google, Inc.

The Arc de Triomphe, seen from street level in Google Street View. © Google, Inc.

The Arc de Triomphe was commissioned in 1806 and completed thirty years later, commemorating those who fought and died for France. Most people on the planet will never go to Paris, never walk around the Arc de Triomphe's plaza, never see the eternal flame at its base.

But the two billion people who have Internet access can see it instantly. Or they can visit base camp on Mt. Everest[45] or swim with sea lions among the Galápagos Islands.[46]

South Base Camp, Khumjung, Mt. Everest, Nepal. © Google, Inc.

Under the sea with Galápagos sea lions, Galápagos Islands. © Google, Inc.

The sweep of our mission has had surprising practical benefits as well. Outside the company, Philip Salesses, Katja Schechtner, and César Hidalgo of the MIT Media Lab compared images of Boston and New York with images of Linz and Salzburg, Austria, to explore what features—dirty streets or the number of streetlamps, for example—made it feel like neighborhoods were rich or poor, and whether those signifiers of economics and class correlated with safety.[47] Eventually, their approach could be used to help cities determine how to allocate scarce resources best: Will neighborhoods feel and become safer if more trees are planted or if roads are repaired?

Google's map products form a platform that more than one million sites and app developers have used to build businesses, ranging from Airbnb to Uber, from Waze to Yelp,[48] serving more than one billion users each week.[vii, 49]

[vii] In fairness, there are also real concerns that Googlers and users raise about privacy as it relates to capturing street-level imagery. We try to be sensitive to these concerns. For example, our default is to blur faces and license plates to help ensure anonymity, as you can see in the image below from the Taj Mahal. At the same time, our algorithms can be a little overzealous and also anonymize our nonhuman friends (see next page).

Taj Mahal, Agra, India. © Google, Inc.

A more traditional mission of creating value for customers or growing profits would never have led us to Street View. And it's a far cry from counting backlinks in order to rank websites. But our broader mission provided the space for Googlers and others to create wonderful things. These bursts of creation and accomplishment were a direct result of articulating Google's mission as something to keep reaching for, just beyond the frontiers of what we can imagine.

The most talented people on the planet want an aspiration that is also inspiring. The challenge for leaders is to craft such a goal. Even at Google, we find that not everyone feels the same strong connection between their work and the company's mission. For example, in our 2013 survey of Googlers, 86 percent of our sales teams agreed strongly that "I see a clear link between my work and Google's objectives," compared to 91 percent for other parts of Google. Same mission. Same company. Different level of connection and motivation. How do you address this?

Adam Grant has an answer. In *Give and Take*, he writes about the power of purpose to improve not just happiness, but also productivity.[50] His answer, like many brilliant insights, seems obvious once it's pointed out. The big surprise is how huge the impact is.

Adam looked at paid employees in a university's fund-raising

A man and his dog, Central Park, New York City. Faces blurred by Google Street View for their privacy. Image discovered by Jen Lin. © Google, Inc.

call center. Their job was to call potential donors and ask for contributions. He divided them into three groups. Group A was the control group, and just did their jobs. Group B read stories from other employees about the personal benefits of the job: learning and money. Group C read stories from scholarship recipients about how the scholarships had changed their lives. Groups A and B saw no difference in performance. Group C, in contrast, grew their weekly pledges by 155 percent (to twenty-three a week from nine a week) and weekly fund-raising by 143 percent (to $3,130 from $1,288).

If reading about someone made such a big difference, Adam wondered, would actually meeting someone have even more of an impact? A group of fund-raisers were given the opportunity to meet a scholarship recipient and ask them questions for five minutes. The result: Over the next month, weekly fund-raising went up by more than 400 percent.

He found this effect persisted in other jobs as well. Lifeguards who read stories about saving drowning swimmers were 21 percent more active in watching over their swimmers. Students editing letters written by other students spent 20 percent more time on them if they first met the authors.[51]

So what is Adam's insight? Having workers meet the people they are helping is the greatest motivator, even if they only meet for a few minutes. It imbues one's work with a significance that transcends careerism or money.

Deep down, every human being wants to find meaning in his or her work. Let's take an extreme example. Is being a fish slicer meaningful work? Chhapte Sherpa Pinasha thinks so. He works at Russ & Daughters, a Manhattan purveyor of smoked fish, bagels, and specialty foods. The forty-year-old started working at Russ & Daughters over a decade ago, but was born in a village in the eastern Himalayas where he lived in a wooden shack, the youngest of four children. At fifteen, he started work, carrying ninety-pound bags of provisions to base camps for Mt. Everest climbers and accompanying foreigners on treks across the mountains. Is his current work less important than helping people summit the tallest peak on the planet? "The two jobs are not really different," he told *New York Times* reporter Corey Kilgannon. "Both involve helping people."[52] Pinasha chooses to see a deeper mission in his work, whereas many people would view the job as "just" slicing lox.

We all want our work to matter. Nothing is a more powerful motivator than to know that you are making a difference in the world. Amy Wrzesniewski of Yale University told me people see their work as just a job ("a necessity that's not a major positive in their lives"), a career (something to "win" or "advance"), or a calling ("a source of enjoyment and fulfillment where you're doing socially useful work").

You would expect that it's easier to consider some occupations a calling than others, but the unexpected discovery is that it's all in how you think about it. Amy looked at doctors and nurses, teachers and librarians, engineers and analysts, managers and secretaries. Across each of these, roughly one-third of people viewed their work as a calling. And people who did so were not just happier, but reported being healthier as well.[53]

Once it's explained, it seems self-evident. But how many of us have taken the time to look for the deeper meaning in our work? How many of our companies make a practice of giving everyone, especially those most remote from the front office, access to your customers so employees can witness the human effect of their labors? Would it be hard to start?

At Google we've begun experimenting with using these personal touches as a way of directly connecting every person with our mission. I spoke recently with a group of three hundred salespeople who spend all day online helping small businesses advertise their wares across the Internet. For the Googlers, the jobs can become rote. But, I told them, those small business owners are reaching out because the problem that is easy for you is hard for them. You've managed hundreds of ad campaigns, but this is their first. When the Paul Bond Boot Company, a custom cowboy boot maker in Nogales, Arizona, wanted to expand beyond word-of-mouth sales, their first ads on Google increased sales by 20 percent. Paul's company was suddenly connected to a much larger world. When we first shared a video of their story with Googlers, they were thrilled and inspired. Nikesh Arora, our SVP of Global Business at the time, called these "magic moments." Watching for them and sharing them keeps Googlers connected to the company's mission. If the benefits from making these connections are half of what Adam has found, they will prove to be superb investments.

If you believe people are good, you must be unafraid to share information with them

Transparency is the second cornerstone of our culture. "Default to open" is a phrase sometimes heard in the open-source technology community. Chris DiBona, leader of Google's open-source efforts, defines it like this: "Assume that all information can be shared with the team, instead of assuming that no information can be shared. Restricting information should be a conscious effort, and you'd better have a good reason for doing so. In open source, it's countercultural to hide information." Google didn't create this concept, but it's safe to say we ran with it.

As an example, consider Google's code base, which is the collection of all the source code—or computer programs—that makes all our products work. This includes the code for almost everything we do, including Search, YouTube, AdWords, and AdSense (those little blue text ads you see on the Internet). Our code base contains the secrets of how Google's algorithms and products work. At a typical software company, a new engineer will be able to see some of the code base for just their product. At Google, a newly hired software engineer gets access to almost all of our code on the first day. Our intranet includes product roadmaps, launch plans, and employee snippets (weekly status reports) alongside employee and team quarterly goals (called OKRs, for "Objectives and Key Results"...I'll talk more about them in chapter 7), so that everyone can see what everyone else is working on. A few weeks into every quarter, our executive chairman, Eric Schmidt, walks the company through the same presentation that the board of directors saw just days before. We share everything, and trust Googlers to keep the information confidential.

At our weekly TGIF all-hands meeting, Larry and Sergey host the entire company (thousands join in person and by video, and tens of thousands watch the rebroadcast online) for updates from the prior week, product demonstrations, welcoming of new hires, and most important, thirty minutes of fielding questions from anyone in the company, on any topic.

The Q&A is the part that matters most.

Everything is up for question and debate, from the trivial ("Larry, now that you're CEO, will you start wearing a suit?" The answer was a definite no), to business-related ("How much does Chromecast cost to make?"), to the technical ("*The Guardian* and *New York Times* today revealed that NSA internal documents claim they are covertly influencing cryptographic products in order to insert vulnerabilities. What can I do as an engineer to help keep our users' data securely encrypted?"), and the ethical ("To me, privacy includes being able to say things online without attaching my real name to them—say, publicly commenting on an Alcoholics Anonymous video on YouTube without outing myself as alcoholic. Does Google still support that kind of privacy?").[viii] Any question is fair game, and every question deserves an answer.

Even the way the questions are chosen is rooted in transparency, using a tool (awkwardly) called Hangouts On Air Q&A. Users can not only submit questions, but also discuss and vote on them. This crowdsourcing prioritizes questions that reflect the interests of an audience.

In 2008, President-elect Obama's transition team used this tool

---

[viii] Questions like this, as well as input from users, led to our decision in 2014 to allow people to use aliases as an alternative to their real names in Google+.

as part of their Open for Questions events, a series of nationwide town halls where everyone in the country was invited to submit questions for the president to answer. Participants submitted more than ten thousand questions and voted more than one million times to determine which were the most important.

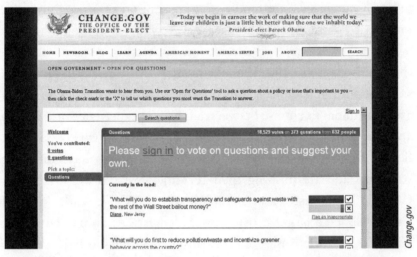

Google Moderator, as used by the Obama-Biden transition team in 2008.

The benefit of so much openness is that everyone in the company knows what's going on. This may sound trivial, but it's not. Large organizations often have groups doing redundant work without knowing it, wasting resources. Information sharing allows everyone to understand the differences in goals across different groups, avoiding internal rivalry. The case against this approach is made by companies that foster internal rivalry and obscure information across teams. Alfred Sloan famously created this type of culture as CEO of General Motors, and it culminated in GM at one point having five major brands, each of which sold cars that competed with the others to varying degrees. For example, Toyota had a single offering in the mid-priced sedan category: the Camry. In the same category, GM offered two cars from Buick (the Allure and the

Lucerne), the Cadillac CTS, two from Chevrolet (the Impala and the Malibu), the Pontiac G8, and the Saturn Aura.[54] Even if one part of GM won a sale, four parts lost.

At Google, we sometimes offer similar products. We minimize unhealthy competition by letting Googlers know about them and explaining why we allow this competition to persist. Often it's to take advantage of "late binding," a corruption of a programming term that we use to indicate that it can be valuable to wait to make a decision. For example, we produce two operating systems: Chrome (primarily laptops and Web browsers) and Android (primarily smartphones and tablets). On one level, it doesn't make sense to ask consumers to choose between a Chrome experience on a laptop and an Android experience on a phone. Both are made by Google: Shouldn't they be the same? But the two teams have different strengths and are pushing their technologies in novel directions. Chrome boots faster and has more robust Wi-Fi, while Android has developed a larger ecosystem of applications on its Play Store. Thus far, the innovation and learning from having both systems outweigh the costs of deciding on one or the other.

We also use an unfortunately named technique common in technology firms called "dogfooding," where Googlers are the first to try new products and provide feedback.[ix] Dogfooders were the first to test-ride in our self-driving cars, supplying valuable feedback on how they work in daily use. This way, Googlers learn what's going on, and teams get valuable, early feedback from real users.

[ix] The expression was popularized in the technology industry by Paul Maritz of Microsoft in a 1988 internal email urging the use of Microsoft's own server product. But executives of Mars Inc., the makers of Kal Kan pet foods, had long been known to literally eat their own dog food. Joel Brenner of *The Independent* reported on July 26, 1992 ("Life on Mars"): "We are standing in the 'cutting room' at Mars's petfood division in Vernon, California. Without hesitation, [VP of sales John] Murray dips his manicured hands into the dog food, plucks a moist brown lump from the thick gravy and pops it into his mouth. 'It's highly palatable and enticing for the animal,' he remarks. 'Really, it tastes just like cold stew.'" In the mid-1990s, I visited the same room and witnessed the same thing.

One of the serendipitous benefits of transparency is that simply by sharing data, performance improves. Dr. Marty Makary, a surgeon at the Johns Hopkins Hospital in Baltimore, Maryland, points to when New York State started requiring hospitals to post death rates from coronary artery bypass surgeries. Over the next four years, deaths from heart surgery fell 41 percent.[55] The simple act of making performance transparent was sufficient to transform patient outcomes.

There are examples of companies that have pushed internal transparency further than we have. Bridgewater Associates, the world's largest hedge fund with $145 billion in assets,[56] takes one such approach: Every meeting is recorded and made available to all employees. Bridgewater's founder, Ray Dalio, explains: "My most important principle is that getting at the truth…is essential for getting better. We get at truth through radical transparency and putting aside our ego barriers in order to explore our mistakes and personal weaknesses so that we can improve."[57]

Recordings are used not just as communication vehicles, but as learning tools. Managers receive periodic compilations that provide important updates about recent events, illustrate how decisions are made, and share how even the most senior people are learning and growing. The recordings are also used to encourage more precise thinking and communication. No more "I never said that" or "That's not what I meant," when it's possible to review what actually happened. A more subtle objective is to reduce politicking. It's hard to go behind someone's back when they can later listen to your meetings.

The value of transparency is paramount to Bridgewater, the foundation of its people philosophy and practices. And it works for them. It's hard to argue with their track record of integrity, strong culture, and decades-long market-beating performance.

At the same time, that level of transparency is further than Google has gone. In part that's because we feel strongly that privacy is an individual right. For example, user data is fiercely protected.

Even when legally required to provide user data in response to law enforcement orders—which we challenge when appropriate—we publish a Transparency Report (www.google.com/transparency report) showing all such requests we can. And when we make mistakes, as we did in 2010 when some Street View cars unintentionally collected payload data from certain unsecured Wi-Fi networks as they drove by, we take steps to fix those mistakes and prevent future ones.

We do find mechanisms to address some of the same issues Bridgewater attacks. The way we solve the "backstabbing" problem, for example, is that if you write a nasty email about someone, you shouldn't be surprised if they are added to the email thread. I remember the first time I complained about somebody in an email and my manager promptly copied that person, which forced us to quickly resolve the issue. It was a stark lesson in the importance of having a direct conversation with my colleagues!

So there are degrees of transparency, of defaulting to open. Most organizations are so far from risk in this area that they have little to lose, and much to gain. Fundamentally, if you're an organization that says "Our people are our greatest asset" (as most do), and you mean it, you must default to open. Otherwise, you're lying to your people and to yourself. You're saying people matter but treating them like they don't. Openness demonstrates to your employees that you believe they are trustworthy and have good judgment. And giving them more context about what is happening (and how and why) will enable them to do their jobs more effectively and contribute in ways a top-down manager couldn't anticipate.

All of us want control over our destinies

Voice is the third cornerstone of Google's culture. Voice means giving employees a real say in how the company is run. Either you believe people are good and you welcome their input, or you don't. For many organizations this is terrifying, but it is the only way to live in adherence to your values.

Many of our people practices originate with our employees. For example, under the US tax code, homosexual couples must pay income tax on the value of health benefits received by a domestic partner, while heterosexual married couples do not have to pay that tax. A Googler emailed Yvonne Agyei, our VP of Benefits, and explained that this wasn't fair. Yvonne's response: "You're right."[58] She implemented a policy that provided extra payments to same-sex couples to cover the additional income tax, becoming one of the first major companies to do so, and the first to do so globally.

Beyond living our values, there are positive benefits from giving employees voice. Ethan Burris of the University of Texas at Austin found that "Getting employees to voice ideas has long been recognized as a key driver of high-quality decisions and organizational effectiveness. Research on voice has shown positive effects of employees speaking up on decision quality, team performance, and organizational performance."[59]

In 2009, Googlers told us through our annual survey that it was becoming harder to get things done. They were right. We had doubled in size, growing to 20,222 employees by the end of 2008 from 10,674 at the end of 2006, and growing to $21.8 billion in revenues from $10.6 billion. But rather than announcing top-down corporate initiatives, our CFO, Patrick Pichette, put the power in Googlers' hands. He launched Bureaucracy Busters, a now-annual program where Googlers identify their biggest frustrations and help fix them. In the first round, Googlers submitted 570 ideas and voted more than 55,000 times. Most of the frustrations came from small, readily addressable issues: The calendar application didn't allow groups to be added, so large meetings took forever to set up; budget approval thresholds were annoyingly low, requiring managers to review even the smallest transactions; time-saving tools were too hard to find (ironic). We implemented the changes Googlers asked for, they were happier, and it actually became easier to do our work.

In contrast, I recall a discussion with an HR leader from one of the ten biggest companies in the country. "Our CEO wants us to be more innovative," she said. "He asked me to call you because Google is known for having an innovative culture. One of his ideas is to set up a 'creativity room' where we have a foosball table, bean-bags, lava lamps, and lots of snacks, so people can come up with crazy ideas. What do you think? How does Google do it?"

I told her a bit about how Google's culture really works, and suggested that perhaps her CEO could try videotaping his staff meetings and sharing the recordings with people so they could see what's going on in the company and what's important to their leaders. I was just floating a crazy idea, but I thought it might be a powerful way to share with employees how decisions were made. I didn't know at the time that Bridgewater was thinking even bigger, taping every meeting. "No," she countered, "we'd never do that." How about having junior people attend leadership team meetings as notetakers, and they could then be vectors for that knowledge across the company? (Jonathan Rosenberg, our former SVP of Product, pioneered this for us.) "No, we couldn't share that information with junior people."

Hmm…okay. How about, when the CEO does employee meetings, seeding the audience with the tough, provocative questions that people are afraid to ask? "Oh no, he would never do that. Think of all the crackpot emails he'd get." A different angle was to have a suggestion box—which she thought might work—and then each quarter let a self-nominated group of employees decide what suggestions to implement. And maybe even give them a budget for it? "Oh no, that won't work. Who knows what they might do?"

This otherwise remarkable company was afraid of giving employees even the tiniest opportunity for direct expression and dialogue with their CEO. At which point I wished her luck with the beanbags and lava lamps.

Culture matters most when it is tested

These three cultural cornerstones—mission, transparency, voice—were at the forefront of our 2010 discussions of how Google should operate in China. While the legal and policy details in China are complicated, the effect is that search engines were forbidden to display results for certain search queries. For example, a search for "Tiananmen Square" would show only results from government-approved sites. This was not making information universally accessible! How could we be true to our values of transparency and voice if we censored our results?

Since 2002, our global website, www.google.com, had periodically become unavailable to users in China.[60] We launched www .google.cn in 2006 to try something different. Since our site was hosted in China, we complied with local laws. When results had to be filtered, other search engines simply censored results, while we added an additional line at the bottom of the screen: "In accordance with local laws, regulations, and policies, some search results have not been displayed." Sometimes the absence of information is information itself.

Chinese Internet users are clever. This small signal was enough to let them know they were being kept in the dark and to seek the truth elsewhere.

We naively hoped that our actions would encourage other companies to provide similar notifications, and that in response it might become unnecessary to censor results. Instead, the opposite happened. At the same time that we started notifying users if their search results were filtered, we noticed that our services would sometimes slow down. Even innocuous search results could take minutes, rather than milliseconds, to show up for users in China, and sometimes our website was completely blocked and unavailable to Chinese users.

Despite this, we continued to grow in China. Users wanted truth.

As the interference with our services worsened, we debated furiously about the right thing to do. Eric had weekly staff meetings of his management team that ran for about two hours, and in many of them we'd spend at least thirty minutes discussing China.

Across the company, Googlers were engaged in the debate as well. These debates happened in product review meetings with teams of engineers, product managers, and senior management, during TGIF, in long email debates that included thousands of people at a time, and in our hallways and cafés.

On the one hand, if our mission meant something and we believed in transparency, how could we be active participants in censorship? If we compromised our culture and principles in this case, where else would we do so? Would we make a stronger statement by not being in China than by collaborating with the government?

On the other hand, China was a country with long social and political cycles; shouldn't we think about change in terms of decades, not years? If we didn't continue engaging on this issue, who would? Wasn't some access to the truth, however limited, better than none?

In 2010, after thousands of hours of discussion and having heard input from employees across the globe, we decided we could not censor our results. Because ignoring the government's censorship directives would be illegal and we of course obey the laws of the countries in which we operate, our only option was to stop providing search on www.google.cn, our site hosted in China.

But we didn't want to turn our backs on our Chinese users. Visitors to our Google.cn page saw a message recommending they visit www.google.hk, our site hosted in Hong Kong. When the British government returned control of Hong Kong to the Chinese in 1997, the terms of the handover were that Hong Kong would be exempt from most mainland Chinese regulations for fifty years. This left us thirty-seven years to be true to our culture and committed to China. The Hong Kong site is often blocked or slowed for mainland users, but it allows us to maintain a local, Chinese-language site for users. Within China, notifying users when search results are censored has since become a common practice. Our presence in search in China has dwindled, but it was the right thing to do.

If you give people freedom, they will amaze you

So, to my astonishment, the phrase "culture eats strategy for breakfast" was pretty spot-on. I realized this only after I'd been at Google five years and was asked to write an article about our culture for *Think Quarterly*.[61] I reflected on our debates as a leadership team and saw that, just as in China, we consistently made decisions based not on economics, but on what supported our values. Time and again, we've let our cultural cornerstones of mission, transparency, and voice anchor us in tackling difficult and divisive issues, debating them, and resolving them into clear strategies: Our culture was shaping our strategy, and not the other way around.

It took me another few years to wonder where the phrase came from. I learned that it was apocryphally attributed to the influential management theorist Peter Drucker.[62] It hangs on the wall of the Ford Motor Company's war room, posted there in 2006 by Ford's president, Mark Fields, as a reminder that a robust culture is essential to success.

If you embark down this path, the road will be bumpy. Culture isn't static. Googlers, for example, have said: "Google's culture is changing and it's just not the company I joined anymore." "I remember when we had just a few hundred people—it was a totally different company. Now we feel like any other big company." "We're just not a fun place anymore."

Each of these quotations is from someone bemoaning that Google has lost its way.

The first quote is from the year 2000 (less than a few hundred employees), the second is from 2006 (six thousand employees), and the last is from 2012 (fifty thousand employees—especially ironic because the word Googlers most used to describe Google's culture in that year was "fun"!) In fact, at every point in Google's history, there has been a sense that the culture was degrading. Almost every Googler longs for the halcyon days of Google's youth…which they tend to define as what Google was like in their first few months. This is a reflection of both how wonderfully inspiring the first few months at Google can be and how quickly Google continues to evolve.

We enjoy a constant paranoia about losing the culture, and a constant, creeping sense of dissatisfaction with the current culture. This is a good sign! This feeling of teetering on the brink of losing our culture causes people to be vigilant about threats to it. I'd be concerned if people stopped worrying.

One way to address this worry is to be open to the discussion and to channel any frustration into efforts to bolster the culture. At Google, we have a secret weapon: Stacy Sullivan. Stacy was hired in 1999 as Google's first Human Resources leader. A champion tennis player, Berkeley grad, and veteran of multiple technology companies, Stacy is smart, creative, bracingly direct, and utterly charming. In short, she's exactly the kind of person Google loves to hire, which makes sense since she helped shape Google's target hiring profile. Today Stacy is Google's chief culture officer, the only person ever to hold that title, charged with making sure Google's culture stays true to itself. She explains: "Even from the first day, we were worried about the culture. It always felt like it was changing, so we have always had to fight to keep the core culture strong."

Stacy built a global network of Culture Clubs, teams of local volunteers charged with maintaining Google's culture in each of our seventy-plus offices. They have modest budgets (typically $1,000 or $2,000 per year) and their brief is to nudge the local office cultures along, staying connected to the rest of Google and encouraging both play and honest discussion. There is no application to be named the leader of a Culture Club. You become one simply by acting like one: taking charge of local office events, being vocal, and—

importantly—emerging as a leader to whom others look for advice on what is "Googley."

Eventually, Stacy finds you and asks you to take on the role.[63]

I mentioned earlier that there are various ways to build great businesses, and companies have been successful with both low-freedom and high-freedom models. Google is clearly in the latter camp. Once you've chosen to think and act like a founder, your next decision is about what kind of culture you want to create. What are the beliefs you have about people, and do you have the courage to treat people the way your beliefs suggest? My personal and professional experience is that if you give people freedom, they will surprise, delight, and amaze you. They will also sometimes disappoint you, but if we were perfect we wouldn't be human. This isn't an indictment of freedom. It's just one of the trade-offs.

The case for finding a compelling mission, being transparent, and giving your people voice is in part a pragmatic one. The growing global cadre of talented, mobile, motivated professionals and entrepreneurs demand these kinds of environments. Over the coming decades the most gifted, hardest-working people on the planet will gravitate to places where they can do meaningful work and help shape the destiny of their organizations. But the case is also a moral one, rooted in the simplest maxim of all: Do unto others as you would have them do unto you.

....

WORK RULES... FOR BUILDING A GREAT CULTURE

☐ Think of your work as a calling, with a mission that matters.

☐ Give people slightly more trust, freedom, and authority than you are comfortable giving them. If you're not nervous, you haven't given them enough.

....

3

Lake Wobegon, Where All the New Hires Are Above Average

Why hiring is the single most important people activity in any organization

Imagine you have won the largest lottery jackpot in US history: $656 million. You can do anything. And improbably enough, you choose to build a championship baseball team.

You have a couple of options. You can take your pile of cash and hire the best players on the planet to play for you. Or you could take the *Bad News Bears* approach: Assemble a ragtag team of misfits and by dint of your coaching, hard work, and deep insight into motivation and human nature, mold them into a team of winners.[x]

Which approach is more likely to give you your championship team? Fortunately, both approaches have been tested.

The first World Series was played in 1903, and there have been 108 World Series since. The New York Yankees have played in forty of them, and won twenty-seven. This is almost four times as many as the St. Louis Cardinals, the next most frequent winners.

A major factor in the Yankees' winning record has been their explicit strategy of paying top dollar for the best players, with a

---

[x] For those of you too young to remember *The Bad News Bears*, think *Major League, The Mighty Ducks, Little Giants, Armageddon, Pitch Perfect*…there's a rich vein of movies to mine for this metaphor.

2013 payroll of $229,000,000.[xi] They have had the highest payroll in baseball every year since 1999 and were the second-highest paying the year before. In fact, 38 percent of the World Series since 1998 were won by one of the two highest-paying teams—either the Yankees or the Boston Red Sox—and 53 percent of the time, one of the two highest-paying teams appeared in the World Series.[64]

This is an extraordinary outcome. If winning were random, then a Major League Baseball team would have a 3 percent chance of winning the World Series each year. So why do the high-paying teams win so often, but not all the time?

It's fairly straightforward to know who the better baseball players are. Their performance is observable, since every game is public and recorded, the rules and positions are well understood, producing a consistent standard of assessment, and their wages are known. And despite the years that have passed since Michael Lewis wrote *Moneyball*, chronicling the Oakland Athletics' clever application of data analytics to player performance, it's still fiendishly difficult to agree on who the absolute best are, or to predict who will have a great year. But it's not hard to identify the top 5 percent or 10 percent of players.

As long as money is no object, a team could hire all the players who performed extremely well last year and have a pretty good chance at fielding a championship baseball team. However, hiring the best players in the league from last year doesn't guarantee that they will be the best players next year. In fact, it's extremely rare.

[xi] Payroll has been a major factor, but not the only one. For example, the Yankees benefited from their influence on the American League in the first half of the twentieth century (see Jeff Katz's *The Kansas City A's and the Wrong Half of the Yankees*, 2007) and aggressive management of their market in the latter half. Tough years to watch for a San Francisco Giants fan. That said, paying for the best performers seems to be linked to success in other sports as well. Upon examining British football between 1996 and 2014, *The Economist* found that "55% of the variation in the number of points scored in any given season can be explained by the amount spent on wages." *The Economist* did allow that this correlation did not prove a causal relationship ("Everything to Play for," *The Economist*, May 10, 2014, page 57).

But you can be pretty confident that their collective performance would at least be in the top half or one-third of all teams in the league.

The downside of this approach, of course, is the cost: Payroll for the Yankees more than tripled to $229,000,000 over this period, rising $163,000,000 since 1998. Today, the sustainability of paying top dollar is being called into question, even by the Yankees. George Steinbrenner was the architect of their strategy of buying the best ballplayers. He has been succeeded by his son, Hal, who planned to reduce the Yankees' 2014 payroll below $189 million to avoid Major League Baseball's luxury tax.[65]

CEOs like to pursue this strategy as well. Marissa Mayer, who had been employee number 20 at Google and was instrumental in shaping our brand and approach to search, became the CEO of Yahoo on July 16, 2012. Over the next year, Yahoo acquired at least nineteen companies,[66] including Jybe (activity and media recommendations), Rondee (free conference calls), Snip.it (news clipping), Summly (news summaries), Tumblr (photo blogs), Xobni (inbox and contact list management), and Ztelic (social-network analysis). The prices for only five of their acquisitions were disclosed, totaling $1,230,000,000 ($1.23 billion). And of the acquisitions listed here, all but Tumblr had some or all of their products shut down once they were acquired, and their people were integrated into Yahoo's existing team.

Buying companies and then shutting down their products is a recent Silicon Valley phenomenon, awkwardly known as acqui-hiring. The ostensible purpose is to obtain people who have demonstrated their capabilities by building great products and who otherwise would not join you as employees.

It's not clear yet whether acqui-hiring is a good way to build successful organizations. First, it's fabulously expensive: Yahoo paid $30 million for Summly, shut it down, and fired all but three employees,[67] retaining only seventeen-year-old founder Nick D'Aloisio and two others. That's $10 million per person. And even when acqui-hires are "cheap," they are still expensive: The thirty-one Xobni employees cost $1.3 million each.[68] And after all that, they still need ongoing salaries, bonuses, and stock awards, just like other employees.

Acqui-hires also see their products killed. That's a painful experience. While the money is supposed to make up for it, I've heard about many acqui-hired engineers across Silicon Valley who are just biding their time until they are fully vested and then plan to strike out on their own again. It's also not clear that acqui-hired employees generate better outcomes than hired employees. Some do, but I haven't seen evidence that it's generally true.

Given that over two-thirds of mergers and acquisitions fail to create value when the products and businesses are kept alive,[69] there would have to be something special about acqui-hired employees to make this strategy work. This isn't to say that acqui-hires are a bad idea. Just that they're not obviously a great idea.

Buying the best seems like the way to go if you're building a baseball team, but it's much trickier if you're building a company. The labor market for employees isn't as transparent as the market for baseball players. The only evidence you have for someone's performance is their resume and what they (and sometimes their references) tell you, rather than actual recordings of them working. Any given position in baseball is pretty much the same across teams; there are only so many different ways to play first base. But there are many different ways to perform a marketing job, for example. And offering higher wages just means you get more applicants, not that you get better applicants or can better sift the great from the mediocre.

For all these reasons, most organizations pursue the *Bad News Bears* strategy, though they don't admit it. What executives will tell you is that they recruit the best people and then groom, train, and coach them into champions. There are three reasons to be skeptical of these claims.

First, if they were really doing this, wouldn't more organizations have champion-level performance? The Yankees are in the World Series 37 percent of the time, and when they play in it, they win 67 percent of the time. There are very few organizations that have that level of performance, and fewer still that have sustained it for a hundred years.

Second, if they were really better at recruiting, shouldn't there be something special about how they recruit? Yet most organizations run recruiting the same way: Post a job, screen resumes, interview some people, pick whom to hire. Nothing more complicated than that. If they are all recruiting the same way, why would any of them get a different outcome than their competitors? And by definition, this means firms are recruiting average talent. Sure they'll get some superstars and some stinkers, just like every other company. But overall, the quality of their new hires is average.

Third, most people simply aren't very good at interviewing. We think we are hiring the best because, after all, aren't we great judges of character? When we start interviews, don't we immediately size up the person and get a pretty good sense of their character and capabilities? And if we never go back and compare our interview notes (if we bothered to take any) with how people actually perform months and years later, so what? We know deep down that we've hired the best.

But we're wrong.

People approach hiring the way Garrison Keillor describes the fictional town of Lake Wobegon, where "all the children are above average." We all think we are great at it, but we never go back to check if we are, and so we never get better. There's ample data showing that most assessment occurs in the first three to five minutes of an interview (or even more quickly),[70] with the remaining time being spent confirming that bias; that interviewers are subconsciously biased toward people like themselves; and that most interview techniques are worthless. Remember when George W. Bush met Vladimir Putin and reported that "I looked the man in the eye....I was able to get a sense of his soul."[71]

In addition to thinking we're superior interviewers, we convince ourselves that the candidate we select is also above average. After all, we wouldn't offer them a job otherwise. But there's a jarring dissonance between the dewy optimism we feel after a great interview and the tepid reality a year later, when we're assessing his or her performance. The handful of stars stick in our memory. We forget our conviction that just about every hire was going to be a star.

So hiring yields average results.

Can't you compensate by training people to be great? Aren't many companies renowned for their leadership academies, global training centers, and tele-learning? Doesn't that let them breed greatness in their new employees?

Not so much. Designing effective training is hard. Really hard. Some experts go so far as to say that 90 percent of training doesn't cause a sustained improvement in performance or change in behavior because it's neither well designed nor well delivered.[72] It's almost impossibly difficult to take an average performer and through training turn them into a superstar. Some may argue that it is nevertheless possible, which is true (and I'll tell you our approach in chapter 9). There are examples of people who were mediocre performers and went on to greatness, though most of those successes are a result of changing the context and type of work, rather than a benefit of training.

Take Albert Einstein, who initially failed to be hired as a teacher and then failed to be promoted at the Swiss Patent Office. He didn't attend a class that transformed him into the best patent clerk that Switzerland had ever seen. Nor did he get a degree in education and start winning teaching awards. His success came because his day job didn't require much of his intellect,[73] so he was free to explore a completely unrelated field.

So we're left with two paths to assembling phenomenal talent. You can find a way to hire the very best, or you can hire average performers and try to turn them into the best. Put bluntly, which of the following situations would you rather be in?

A. We hire 90th percentile performers, who start doing great work right away.

B. We hire average performers, and through our training programs hope eventually to turn them into 90th percentile performers.

Doesn't seem like a hard choice when it's put that way, especially once you realize there's probably enough money in your budget to get these exceptional people—it's just being spent in the wrong places. Companies continue to invest substantially more in training than in hiring, according to the Corporate Executive Board.[74]

	Training spend	Hiring spend
Per employee	$606.36	$456.44
% of total HR expense	18.3%	13.6%
% of revenue	0.18%	0.15%

Companies spent more on training current employees than on hiring new employees. Data from 2012.

Companies then turn vice into virtue by bragging about how much they spend on training. But since when is spending a measure of quality results? Do people boast, "I'm in great shape—I spent $500 on my gym membership this month?" The presence of a huge training budget is not evidence that you're investing in your people. It's evidence that you failed to hire the right people to begin with. In chapter 9 I'll suggest some tactics to reduce your training budget, which you can then repurpose for hiring.

At Google, we front-load our people investment. This means the majority of our time and money spent on people is invested in attracting, assessing, and cultivating new hires. We spend more than twice as much on recruiting, as a percentage of our people budget, as an average company. If we are better able to select people up front, that means we have less work to do with them once they are hired. The worst case with a 90th percentile candidate is that they have an average year. They are unlikely to become the worst performer in the company. An average candidate, however, will not only consume massive training resources, but is also just as likely to end up performing well below average as above average.

Why did we decide to front-load our people investment by focusing on an unorthodox approach to hiring?

We had no choice.

Google started with two guys in a dorm room, in a crowded market where users could abandon us for competitors with just a single mouse click. Since the beginning, we knew that the only way to compete would be to have the most accurate, fastest search product in the world, but thought we would never have enough engineers to build what we needed: Web crawlers to identify and categorize everything on the Internet, algorithms to sift meaning from what was out there, tools to translate more than eighty languages, tests to make sure everything actually worked, data centers to host and serve all this data, and eventually hundreds of other products that needed development and support. Our greatest single constraint on growth has always, always been our ability to find great people.

For many years, we didn't have the huge advantage that the Yankees did: money. Simply buying the best wasn't an option for Google in its formative years, just as it isn't an option for most organizations. In 1998, Google had no revenues and for years paid among the lowest salaries in the industry. As late as 2010, most people who joined Google still took significant salary cuts when joining, some of 50 percent or more. Convincing people to give up their salaries and join this crazy little start-up was no easy task. Like many others, I took a pay cut to join Google, and still remember the words from my division CEO at GE on my last day: "Laszlo, this Google thing sounds like a cute little company. I wish you luck, but when it doesn't work out, give me a call and we'll have a job for you."

Google was also late to the search game, as Yahoo, Excite, Infoseek, Lycos, AltaVista, AOL, and Microsoft were already major players. We had to impress and inspire candidates, and convince them that Google had something special to offer. But even before we could persuade people to join, we had to figure out a new way to hire people, to ensure we had a better hiring result than other companies.

Sifting the exceptional from the rest required radically rethinking hiring, and I'll detail exactly how we did it in the next two chapters. The good news is that it doesn't have to cost more money, but you do have to make two big changes to how you think about hiring.

The first change is to hire more slowly.

Only 10 percent of your applicants (at best!) will be top performers, so you go through far more applicants and interviews. I say at best, because in fact the top performers in most industries aren't actually looking for work, precisely because they are top performers who are enjoying their success right where they are. So your odds of hiring a great person based on inbound applications are low.

But it's worth the wait because, as Alan Eustace, our SVP of Knowledge, often says, "A top-notch engineer is worth three hundred times or more than an average engineer....I'd rather lose an entire incoming class of engineering graduates than one exceptional technologist." [75]

One such is Jeff Dean, an early Googler and key mind behind the search algorithms that enable the fastest, most accurate search on the planet. Jeff, in collaboration with a handful of others, completely reinvented our approach to search multiple times. For example, in the early days, Jeff, Sanjay Ghemawat, and Ben Smith figured out how to keep our search index in memory, rather than served from discs. That by itself was a threefold improvement in efficiency.

Jeff's also a terrific guy and is held in incredibly high esteem by his colleagues. There's an internal site where Googlers submit "facts" about Jeff, similar to achievements attributed to Dos Equis' Most Interesting Man in the World:

- Jeff Dean's keyboard doesn't have a Ctrl key because Jeff Dean is always in control.
- When Alexander Graham Bell invented the telephone, he saw a missed call from Jeff Dean.
- Once, when the index servers went down, Jeff Dean answered user queries manually for two hours. Evals showed a quality improvement of 5 points.
- In 1998, scientists added a leap second to December 31st to give Jeff Dean time to fix the Y2K bug—in every system known to man.
- Jeff Dean once described Zeno's paradox to an innocent bystander. The bystander never moved again.
- To Jeff Dean, "NP"[xii] means "No Problemo."
- If the world was on the brink of global thermonuclear war, they would have WOPR play against Jeff rather than itself.
- Newton once said, "If I have seen far, it is because Jeff Dean will stand on my shoulders."

[xii] I ran the abbreviation "NP" by my close friend Gus Mattammal, who has degrees in math, physics, and business and is director of Advantage Testing of Silicon Valley, an elite tutoring and test preparation firm. I figured if anyone could explain NP, he could. Gus told me, "Class NP contains all computational problems such that the corresponding decision problem can be solved in a polynomial time by a nondeterministic Turing machine." Ummm....He then translated for me: "Unless you're a computer scientist, 'NP problems' can just be used to stand for 'really, really hard problems to solve.'"

As special as Jeff is to us, he's not alone. Salar Kamangar had the insight on how to create auctions for search terms, and worked closely with engineer Eric Veach to build our first ad systems. In publishing, for example, magazines will list a price they charge advertisers for every thousand readers. Instead of naming a price up front, Salar dreamed up running an auction for every word or phrase a user might search for. Google doesn't arbitrarily decide in what order ads are presented. Rather, our advertisers bid for the position they want in the list of ads, which can cost from less than a penny to more than $10 per word. These insights translated directly into billions of dollars of value for our shareholders, hundreds of billions of dollars in new business for our advertisers, and happier users who could now find exactly what they wanted across the entire Web.

Other exceptional Googlers include Diane Tang—one of only a handful of engineers to earn the accolade of Google Fellow, an honorific reserved only for those who have had the greatest technical contributions—who for years led the team focused on making sure ad quality continued to improve and recently took on a confidential project at Google[x]. Dr. Hal Varian, who literally wrote the book on microeconomics, leads our economics team. Charlotte Monico, a London-based member of our people operations team, is one of over a dozen Googlers to have taken part in the Olympic games. Vint Cerf, known as "the co-father of the Internet" for his seminal work co-inventing the Internet, is our lead evangelist. The inventor of the optical mouse (Dick Lyon) and founders or cofounders of Excite (Joe Kraus and Graham Spencer), Ushahidi (a crowdsourcing utility that allows citizen journalists and eyewitnesses to report violence in Africa, created by Ory Okolloh), Chrome (Sundar Pichai and Linus Upson), and Digg (Kevin Rose) work alongside one another and tens of thousands of other remarkable people.

How can you tell if you have found someone exceptional? My

simple rule of thumb—and the second big change to make in how you hire—is: "Only hire people who are better than you."

Every person I've hired is better than me in some meaningful way. For example, Prasad Setty, VP of People Analytics and Compensation, is more analytically insightful. Karen May, our VP of People Development, is a more thoughtful counselor, in part because her emotional intelligence is much higher than mine. Nancy Lee, who leads diversity and youth education programs for the company, has a fearlessness and clarity of vision that I envy. Sunil Chandra, VP of Staffing and People Services, is more operationally disciplined and insightful, and seems able to make any process faster, cheaper, and better for users. Any of these people could do my job tomorrow. I learn from them every week. And I waited a long time to hire each one. Karen turned me down for four years before I eventually hired her. It takes longer to find these exceptional people, but it's always worth the wait.

In addition to being willing to take longer, to wait for someone better than you, you also need managers to give up power when it comes to hiring. I should disclose up front that newly hired managers at Google hate this! Managers want to pick their own teams. But even the best-intentioned managers compromise their standards as searches drag on. In most companies, for example, they set very high bars for the quality of administrative assistants they want on the first day of a search, but by day ninety most managers will take anyone who will answer a phone. Even worse, individual managers can be biased: They want to hire a friend or take on an intern as a favor to an executive or a big client. Finally, letting managers make hiring decisions gives them too much power over the people on their teams. (I'll discuss in later chapters why we actually work to minimize managers' power.)

After six months or so, our new managers see that the quality of people they are hiring is better than they have experienced anywhere in the past, and that they are surrounded by remarkable people who also made it through the same rigorous process. I wouldn't say they come to love not making hiring decisions, but they appreciate it.

One of the delightful side effects of this rigor is that the best people don't always look like what you'd expect. When Google was small and hiring just a few hundred people a year, it was easy and efficient to hire only people with sterling pedigrees: graduates of Stanford, Harvard, MIT, and similar schools who had worked at only the most highly regarded companies. As we grew to need thousands of new employees each year, we learned that many of the best people didn't go to those schools. Not shocking to you, perhaps, but these were early days at Google and, quite frankly, our approach was more elitist then. We were still managing people issues based on our best instincts, which could be just as flawed as anyone else's, instead of complementing them with data.

So we started seeking out candidates who had shown resilience and an ability to overcome hardship. We now prefer to take a bright, hardworking student who graduated from the top of her class at a state school over an average or even above-average Ivy League grad. The pedigree of your college education matters far less than what you have accomplished. For some roles, it's not important whether you went at all. What matters is what you bring to the company and how you've distinguished yourself. Which in a way is as it should be, considering that one of our founders never finished his university education either. Though we now recruit computer scientists from over three hundred schools in the United States and more from all over the world, some of our best performers never set foot on a college campus.

But I'd be negligent if I didn't close with a word of caution. In the wake of Enron's 2001 meltdown, Malcolm Gladwell wrote a piece for *The New Yorker* titled "The Talent Myth: Are Smart

People Overrated?" skewering both Enron and McKinsey for their obsession with "smart people": "The broader failing of McKinsey and its acolytes at Enron is their assumption that an organization's intelligence is simply a function of the intelligence of its employees. They believe in stars, because they don't believe in systems." [76]

While that didn't quite square with my own experience of McKinsey, which had a robust set of internal systems for people development and counseled clients to have the same, I agree that blindly hiring for brains and giving them unbounded freedom to do what they will is a recipe for sudden and catastrophic failure. You obviously want to hire the best people, but "best" isn't defined by a single attribute like intelligence or expertise.

As I'll share in chapter 8, being a star in one environment doesn't make you a star in a new one. So making sure someone will thrive in your environment becomes critical. In chapter 5 I'll detail how we do this at Google by looking for a wide range of attributes, among the most important of which are humility and conscientiousness. These other attributes are such significant factors in our hiring process that they caused Ben Gomes, who was involved with two of Google's first three patents related to search, to observe: "It's an interesting phenomenon: interviewing people better than you and saying 'No.' "

The lesson of "The Talent Myth" was not "Don't hire smart people." It was "Don't hire exclusively for smarts." Sound advice. Superb hiring isn't just about recruiting the biggest name, top salesperson, or cleverest engineer. It's about finding the very best people who will be successful in the context of your organization, and who will make everyone around them more successful.

Hiring is the most important people function you have, and most of us aren't as good at it as we think. Refocusing your resources on hiring better will have a higher return than almost any training program you can develop.

....

WORK RULES...FOR HIRING (THE SHORT VERSION)

☐ Given limited resources, invest your HR dollars first in recruiting.
☐ Hire only the best by taking your time, hiring only people who are better than you in some meaningful way, and not letting managers make hiring decisions for their own teams.

....

4

....

Searching for the Best

The evolution of Google's "self-replicating
talent machine"

As we wrapped up a Google board meeting, Paul Otellini, the CEO of Intel and a board member, concluded: "What's most impressive is that your team has built the world's first self-replicating talent machine. You've created a system that not only hires remarkable people, but also scales with the company and gets better with every generation." I felt like a marathoner collapsing with relief across the finish line. It was April 2013 and Google had added more than ten thousand people in the past two years.

In fact, we've grown by about five thousand people almost every year. To get there, we start with the 1,000,000 to 3,000,000 people who apply for jobs each year, which means we hire about 0.25 percent of the people we consider. As a point of comparison, Harvard University in 2012 extended offers to 6.1 percent of its applicants (2,076 admitted out of 34,303 applicants). It's a very hard place to get into, but almost twenty-five times easier than getting hired by Google.

It really did start with the founders

Larry and Sergey, with input from Urs Hölzle (one of our first ten hires and now our SVP of Technical Infrastructure), laid the foundation for Google's hiring system. It started with a desire to hire only the smartest people. Later we refined the process because IQ alone doesn't make someone creative or a team player, but it was a great starting point.

As Urs explained, "I really had a bad experience where I was working in a small start-up, seven people, and we were acquired by Sun, and the team grew from seven to, like, fifty very quickly, and our productivity was less than before. Because few of the new engineers were of the same quality. And they were costing us more time than they were providing, and we would have been much better off with a team of fifteen if, you know, everyone was really very good. I was sort of afraid of having Google with fifty engineers be less productive than Google with ten engineers."

The founders realized it was important to hire by committee, often interviewing candidates together while sitting around the ping-pong table, which doubled as our only conference table. They intuited that no individual interviewer will get it right every time, an instinct that would later be formalized in our "wisdom of the crowds" study in 2007, which we'll discuss shortly. Even Susan Wojcicki, who knew Larry and Sergey well and owned the garage they were renting as Google's first office, had to interview for her original job as our first marketing leader.

Importantly, they also had the instinct to hew to an objective standard, ideally enforced by having a single, final, central reviewer who is charged with upholding that standard. Today we split that responsibility across two teams of senior leaders, one for product-management and engineering roles and another for sales, finance, and all other roles. And we have one final reviewer of every—yes, every—candidate: our CEO, Larry Page.

The sole purpose of these two teams is to ensure that we stay true to the high-quality bar set by the founders. If you start a company or team, you know exactly what you are looking for in a new hire: someone just as motivated, clever, interesting, and passionate as you are about the new venture. And the first few people you hire will meet that standard. But they in turn won't uniformly hire to the same standard as you, not because they are bad or incompetent people, but because they won't have precisely the same understanding of what you are looking for.

Each generation of hiring will therefore be a slightly poorer version of the hiring done by the prior generation. As you get bigger, there will also be more temptation to hire a friend or customer's child to help them out or build the relationship. These are almost always a compromise of quality. The result is that you go from hiring stellar people as a small company or team to hiring average people as a big company.

The early days: hiring astounding people at a snail's pace

Before 2006, Googlers tried anything and everything to find candidates. We tested traditional tactics like advertising jobs on websites like Monster.com. These worked, but not well. For every candidate we hired, there were tens of thousands that we didn't.[xiii] Hours and hours were spent sifting through this flood of applications.

Like everyone else, we did reference checks, but we also built an applicant tracking system that would check a candidate's resume against the resumes of existing Googlers. If there was overlap—say you went to the same school in the same years as a Googler, or worked at Microsoft at the same time—the Googler would often get an automated email asking if they knew you and what they thought of you. The idea was that since references the candidate provides are almost always glowing, these "backdoor" references, we thought,

---

[xiii] In the spring of 2012, we started deploying algorithms to better match candidates with jobs. By mid-2013, hiring yield had increased 28 percent (that is, for every 1,000 applicants, we are hiring 28 percent more people than in the past).

would be more honest. And this approach would screen out people who "kiss up and kick down."[xiv]

All this information and more would be assembled into a hiring packet of fifty pages or more per candidate and reviewed by a hiring committee. There were many hiring committees, and each would be composed of people who were familiar with the job being filled but didn't have a direct stake in it. For example, a hiring committee for online sales roles would be made up of salespeople, but would not include the hiring manager or anyone who would directly work with the candidate. This was to ensure objectivity.

We contracted with recruiting firms. But it was difficult for them to understand what we were looking for, since we wanted to hire "smart generalists" rather than experts. The firms were mystified that we'd prefer hiring someone who was clever and curious over someone who actually knew what he was doing. Their confusion gave way to frustration when we insisted on paying only for successful hires rather than providing retainer fees, as most clients did. And that wasn't all. We required dozens of interviews, rejected more than 99 percent of candidates, and typically offered candidates lower pay than they were currently getting.

We tried crazy things. In 2004 we ran a billboard in Cambridge, Massachusetts, and off the 101 Freeway in California, with a cryptic puzzle on it, hoping that curious and ambitious computer scientists would solve it. This was the billboard:

[xiv] The original source for this phrase appears to be Roos Vonk, a professor at Radboud University in the Netherlands. In her wonderfully titled 1998 paper "The Slime Effect: Suspicion and Dislike of Likeable Behavior Toward Superiors" (*Journal of Personality and Social Psychology* 74, no. 4 (1998): 849–864) she appropriated the Dutch phrase "licking upward–kicking downward." Her paper described several experiments, the first of which had the goal of showing that people who behave this way are "(a) extremely dislikeable and (b) highly slimy."

The cryptic billboard.[77]

Correctly solving this puzzle[xv] led you to a Web page, with a second puzzle:

Congratulations. You've made it to level 2. Go to **www.Linux.org** and enter *Bobsyouruncle* as the login and the answer to this equation as the password.

$$f(1) = 7182818284$$
$$f(2) = 8182845904$$
$$f(3) = 8747135266$$
$$f(4) = 7427466391$$
$$f(5) = \underline{\hspace{2cm}}$$

A second puzzle. © Google, Inc.

[xv] In case you were wondering, the answer is 7,427,466,391.

If you were able to solve this second puzzle,[xvi] you were shown the following:

Congratulations.

Nice work. Well done. Mazel tov. You've made it to Google Labs and we're glad you're here.

One thing we learned while building Google is that it's easier to find what you're looking for if it comes looking for you. What we're looking for are the best engineers in the world. And here you are.

As you can imagine, we get many, many resumes every day, so we developed this little process to increase the signal to noise ratio. We apologize for taking so much of your time just to ask you to consider working with us. We hope you'll feel it was worthwhile when you look at some of the interesting projects we're developing right now. You'll find some links to more information about our efforts below, but before you get immersed in machine learning and genetic algorithms, please send your resume to us at problem-solving@google.com.

We're tackling a lot of engineering challenges that may not actually be solvable. If they are, they'll change a lot of things. If they're not, well, it will be fun to try anyway. We could use your big, magnificent brain to help us find out.

Some information about our current projects:

- Why you should work at Google
- Looking for interesting work that matters to millions of people?
- http://labs.google.com

©2004 Google

The reward for solving both puzzles. © Google, Inc.

The result? We hired exactly zero people.[xvii] The billboard generated a lot of press, but it was a waste of resources: The staffing team had to deal with a flood of resumes and inquiries. Most visitors didn't make it through both puzzles. In interviewing those who did, we learned that doing well in solo competitions doesn't always translate into being a team player. And while people who win these contests can be brilliant, it's often only in one field. Or they are accustomed to solving problems with finite ends and clear solutions,

---

[xvi] These are 10-digit sequences in e that sum to 49. f(5) is 6819025515.

[xvii] In 2013, we checked our hiring records to see if this was still true. While we didn't hire anyone directly as a result of the billboard, twenty-five current Googlers did say that they'd at least seen the billboards. They all thought it was a fun promotion but, as one pointed out: "Puzzles are great, but having a long cryptic text on a billboard along the freeway may not be the best way to advertise." Or make the freeways any safer!

rather than navigating the complexity of real-world challenges. That's an issue at Google, where we look for people who can not only solve today's problems, but can also solve whatever unknown problems may come up in the future.

During our hiring process, we gathered many perspectives on each candidate, in the belief that any single view might be skewed, but some of what we collected was irrelevant. Every candidate had to provide an SAT score, graduate school test scores if they had them, and college transcripts. When I interviewed, I couldn't believe that Google wanted me to call my college and get transcripts from thirteen years ago. The requirement sounded even nuttier to people who had been out of school for twenty or thirty years.[xviii]

We thought requesting grades and transcripts was a blunt instrument to get at smarts. And it did weed out the disappointing number of people who lied about their records. But in 2010, our analyses revealed that academic performance didn't predict job performance beyond the first two or three years after college, so we stopped requiring grades and transcripts except from recent graduates.

In the mid-2000s, interviewers could ask candidates any questions they wanted, but they didn't follow any particular structure, so their feedback often lacked insight. The absence of coordination across interviewers also meant we often forgot to ask about some specific attribute, so the candidate had to come back for yet more interviews.

This made for a miserable experience for many candidates. The press at the time was filled with horror stories about the Google hiring process: "They treat you like a disposable, expendable object";[78] "I am sad to say that reports of the company (or its recruiting organ) being arrogant [and] rude...are NOT exaggerated."[79]

[xviii] Even though we'd long requested grades and transcripts, we were always mindful of the limitations I describe in chapter 5. We have always tried to get as good a sense of the whole applicant as possible.

As you might imagine, the hiring machine moved glacially. Being hired by Google could take six months or longer, and a candidate could endure fifteen or even twenty-five interviews before getting an offer. A Googler might interview ten or more candidates out of hundreds or thousands who applied for a single job, investing ten to twenty hours of time in interviewing and writing feedback for every hire that was eventually made. Now multiply that across the fifteen to twenty-five interviews each successful candidate went through, and you have 150 to 500 hours of employee time invested in every hire before even considering the time spent by recruiters, hiring committees, senior leaders, and the founders.

But in retrospect, this was the right trade-off at the time. The hiring machine was overly conservative by design. It focused on avoiding false positives—the people who looked good in the interview process but actually would not perform well—because we would rather have missed hiring two great performers if it meant we would also avoid hiring a lousy one. A small company can't afford to hire someone who turns out to be awful. Bad performers and political people have a toxic effect on an entire team and require substantial management time to coach or exit. Google was growing too quickly and had too much at stake to risk that. So we kept roles open until we found exactly the right candidate. As Eric Schmidt once told me, "The reality is, there are some employees you should get rid of, but the goal of recruitment should be to have no such employees!"

And as we'd hoped, the combination of our stringent hiring bar and exhaustive focus on recruiting meant we were successfully hiring remarkable people. Among the first hundred hires were people who would go on to be CEOs (of Yahoo and AOL), venture capitalists, philanthropists, and, of course, continue on as Googlers and lead some of Google's most important initiatives. Susan Wojcicki, for example, led our advertising product efforts before moving over to lead YouTube.

In fact, sixteen years later, about one-third of the original hundred hires are still at Google.[xix] It's rare among start-ups for early hires to persist this long, and even rarer for them to be able to continue growing personally and professionally as the company scales from tens of people to tens of thousands.

One of the main reasons we focus so much on growing the company is to have enough great jobs for our people. Larry once explained: "We're a medium-size company in terms of employee count. We have tens of thousands of employees. There are organizations out there that have millions of employees. That's a factor of a hundred, basically. So imagine what we could do if we had a hundred times as many employees." He often tells employees that in the future they could each be running a business as big as Google is today, while still being part of the company.

So the hiring system was functional, but it was a far cry from being a self-replicating talent machine. By the time I joined in 2006, it felt like every other person I met in Silicon Valley had had some painful experience with Google's hiring juggernaut: One software engineer I met told me about the arrogance of the Googlers who interviewed him; my real estate agent's brother had been rejected by Google and then a week later gotten a call from a Google recruiter for the same job; the waiter at the local diner had a friend who was interviewing at Google—and had been interviewing for the past eight months! Even Googlers complained about how long and arbitrary the hiring process seemed, even though they all agreed that it did result in remarkable hiring quality.

It was obvious that we had a problem. If each hire took 250 hours

---

[xix] Among these are Salar Kamangar, Urs Hölzle, Jeff Dean and Sanjay Ghemawat (both VPs and Google Senior Fellows), Jen Fitzpatrick, Ben Smith, and Ben Gomes (all VPs of engineering), Stacy Sullivan (VP and chief culture officer), Matt Cutts (head of our webspam team, and one of the most outspoken, public, and crisp thinkers on issues related to Google), Miz McGrath (a leader in overseeing the quality of ads), Krishna Bharat (creator of Google News and the founder of our Bangalore site), and many others.

of employee time, and we wanted to hire just 1,000 people each year, we'd need to invest 250,000 hours of time. Put another way, it would take 125 people working full-time to hire 1,000 people. And before 2007 we didn't even have hiring goals. Our mandate was to hire as many bright people as we could, so we kept adding hundreds and hundreds of recruiters and taking more and more Googler time. Our hiring process was simply too resource intensive, too time consuming, and too painful for candidates.

Needles in a very big haystack: finding the best candidates among the seven billion people out there

In the early days and for many years, our best source of candidates was referrals from existing employees. At one point, more than half of all our hires were referrals from other employees. In 2009, however, we saw the rate at which employees were making referrals start to decline. Since referrals had been our top source of hires for our first ten years, this was pretty alarming.

The easiest and most obvious response was to increase the reward for successful referrals. The logic was that if the average Googler would make seven referrals for a chance at being paid a $2,000 referral bonus, surely they would suggest even more names for more money. We increased the bonus to $4,000.

And it didn't change our referral rate at all.

It turned out that nobody was meaningfully motivated by the referral bonus. When I asked Googlers why they referred their friends and colleagues to Google, I was floored by the strength of their responses:

> "Are you kidding? This place is the best! I'd love for my friends to work here."
> "The people here are so cool. I know someone who would fit right in."

"I get to be part of something bigger than me. How many
people can say that?"

My first reaction was "How much Kool-Aid have these people
been drinking?" But as I spoke to more people and looked at our
surveys, I saw these responses weren't a fluke. People actually loved
their work experience and wanted other people to share it. Only
rarely did people mention the referral bonus.

A referral bonus is an extrinsic motivator, meaning that it is
motivation that comes from outside yourself. Other extrinsic motivators include public recognition, salary increases, promotions, tro-
phies, and trips. This is in contrast to intrinsic motivators, which
come from inside yourself. Examples of these include a desire to
give back to your family or community, slaking your curiosity, or
the sense of accomplishment or pride that comes from completing a
difficult task.

What we learned is that Googlers were making referrals for
intrinsic reasons. We could have offered $10,000 for each referral
and it likely wouldn't have made a difference.

But something didn't make sense. If people were making referrals
for intrinsic reasons, why did the referral rate slow down? Were people having less fun at Google? Were we straying from our mission?

No. We were just doing a really poor job of managing our referrals. Even though referrals had a higher yield than any other source
of candidates, meaning that a candidate coming through a referral
was more likely to be hired than a candidate coming through a website or recruiting firm, we still hired far less than 5 percent of the
people who were referred to us. This was frustrating for Googlers.
Why keep referring good people if less than one in twenty actually
got hired? And even worse, candidates were suffering through far
too many interviews, and referrers weren't being kept in the loop
about what was happening to their friends.

To address these issues, we drastically reduced the number of interviews each candidate went through. We also developed a white-glove service for referrals, where referred candidates get a call within forty-eight hours and the referring Googler is provided weekly updates on the status of their candidates. Googlers and candidates were happier with the process, but the number of referred candidates didn't change. We still hadn't solved the mystery of why we were getting fewer referrals in the first place.

Oops—our employees don't know everyone in the world

Our overreliance on referrals had simply started to exhaust Googlers' networks. In response, we started introducing "aided recall" exercises. Aided recall is a marketing research technique where subjects are shown an ad or told the name of a product and asked if they remember being exposed to it. For example, you might be asked if you remember seeing any laundry detergent commercials in the past month. And then you might be asked if you remember seeing any Tide commercials. A little nudge like that always improves people's recollections.

In the context of generating referrals, people tend to have a few people who are top of mind. But they rarely do an exhaustive review of all the people they know (though one Googler referred her mother—who was hired!), nor do they have perfect knowledge of all the open jobs available. We increased the volume of referrals by more than one-third by jogging people's memories just as marketers do. For example, we asked Googlers whom they would recommend for specific roles: "Who is the best finance person you ever worked with?" "Who is the best developer in the Ruby programming language?" We also gathered Googlers in groups of twenty or thirty for Sourcing Jams. We asked them to go methodically through all of their Google+, Facebook, and LinkedIn contacts, with recruiters on standby to follow up immediately with great candidates they suggested. Breaking down a huge question ("Do you know anyone we should hire?") into lots of small, manageable ones ("Do you know anyone who would be a good salesperson in New York?") garners us more, higher-quality referrals.

But these efforts weren't enough to meet our voracious hiring needs. Even with a hire rate more than ten times better than our average hiring yield, we would need more than 300,000 referrals every year to grow as fast as we wanted. In our best year we received fewer than 100,000.

Along the way, we noticed something startling. The very best people aren't out there looking for work. Great-performing people are happy and being amply rewarded where they are today. They don't occur to people as referrals, because why would you bother referring someone who is happy at their current job? And they certainly don't apply for jobs.

So we rebuilt our staffing team. In the past they had focused on filtering what came in: screening resumes and scheduling interviews. Now they've become an in-house recruiting firm, with the goal of seeking out and cultivating the best people on the planet. Using a homegrown product called gHire, a candidate database we've built and enhanced with a variety of tools for sifting through and tracking candidates, hundreds of brilliant recruiters find and cultivate these individuals over time—sometimes over years.

The result is that our in-house search firm finds more than half of our hires each year, at a cost far lower than using outside firms, with deeper insight into the market, and while providing candidates with a warmer, more intimate experience.

And each year, technology makes it easier to find great people. Thanks to (of course) Google Search and sites like LinkedIn, finding people at different companies is fairly straightforward. In fact, it is now possible to identify virtually every person working in a particular company or industry, and from there decide whom to recruit. We call this a "Knowable Universe" exercise: systematically locating every person within a universe of job types, companies, or candidate profiles.

Want to know everyone who graduated from Cornell University? In mid-2013 I typed "Cornell" into LinkedIn and got a list of 216,173 people in less than a second. And if someone on the team happened to have attended Cornell, they would have access to the entire alumni database. Whether you are interested in recruiting from a single school, company, or professional or personal background, it's straightforward to generate lists of hundreds or thousands of potential candidates.

Even information that an individual may have put on the Internet and then deleted can sometimes still be found. The Wayback Machine, a service of the Internet Archive, regularly makes backups of more than 240 billion Web pages and has searchable records going back to 1996. We use the Wayback Machine only if we think it might help the candidate. For example, we had an applicant who had started a website in 2008 (great!) that had since been acquired (terrific!), but the current site had a heavily misogynistic tone (uh-oh) that was outside the bounds of even our expansive view of free speech. The candidate was going to be rejected, but because I've seen that most people who make it this far in our process are good people, I suggested checking earlier versions of the site. We saw that his original site was more like a college newspaper, covering sports, movies, and celebrities, and only after his company was acquired and he had left had the content shifted. We hired him.

There are a handful of outstanding companies from which, using these techniques, we have assembled a list of essentially all their employees and made an (admittedly imperfect) assessment of who would be a good fit at Google. Obvious candidates for this research are other large technology firms. I'm sure they've made similar lists of our people (and if they haven't, I've no doubt they now will). Once we generate a list, we review portions of it with

Googlers who have expertise in that area or may know those individuals. We spot-check online to see if there's anything else to help us identify those who would be most successful at Google. And then we reach out to network and build a relationship. This might mean an email or call, or even meeting at a conference. Recruiters typically start the relationship, but sometimes the best contact is one of our engineers or executives. And while there may not be an opportunity today, it's always possible the candidate has a bad day a year later and remembers that great conversation they had with a Google recruiter. Jeff Huber, our longtime SVP of engineering for Ads and Apps and now a part of Google[x], where he works on Google's next big bets, personally recruited more than twenty-five senior engineers, one of whom he'd cultivated for ten years across three companies before finally convincing her to join him at Google.

Today our own Google Careers website is one of our best sources of candidates, though we're hard at work making it even better. Corporate job sites are awful. They are difficult to search, filled with generic job descriptions that don't tell you anything about what the job really is or what the team you'll be part of is like, and provide no feedback on whether you'd be good for a role or not. We started addressing this in 2012. For example, a candidate can now submit not only a resume but also develop a personal skill profile. Using Google+ "circles" (a group you select and which has access only to what you want to share... no more sharing the pics of your friend's wild bachelor party with your boss), they can choose to share those skills with Google, other employers, or any subset of people and organizations they wish. They can get in touch with current Googlers to find out what it's really like to work here. With an applicant's permission, we're then able to stay in touch with people who have skills we don't need today but may in the future, and reach out to them as our needs change.

We don't use many recruitment firms, not because they are bad (some of my best friends are executive recruiters... really) but because the stringency and peculiarity of our hiring standards and process, combined with our existing capabilities in recruitment, mean that there are only a handful of situations where recruitment firms help us. And in those cases, they are invaluable. For example, there are countries where our presence is small and we simply don't know the local talent pool. Our Korean office has more than a hundred people, in a country where we are far behind (but gaining on!) Naver, the local Internet portal and search engine. Most people work for one of the national *chaebol*, family-owned conglomerates, and Google is very much an upstart. For senior hires, an introduction from a trusted recruiting partner is invaluable.

And there are sometimes incredibly confidential and sensitive searches where the professionalism of a recruitment firm is very helpful in reaching out to candidates whose jobs may be at risk if it were known they were even entertaining a discussion with another company. A handful of firms have been particularly helpful over the years, but we've found that more important than the firm is the quality of the individual search professionals working with us. In other words, there's more variance in quality within search firms than across search firms, so selecting the individual search consultants you work with is more crucial than selecting the company.

The last source of candidates that we, and most organizations, have used is job boards. These are websites where for a fee an employer can post a job and then receive a flood of applicants. Popular examples are Monster, CareerBuilder, Dice, and Indeed. The Google experience has been that job boards generate many, many applicants and vanishingly few actual hires. Our hypothesis is that, since Google is fairly well known today, a more motivated candidate will show the modest initiative required to actually go to Google Careers and apply directly. A less motivated candidate will apply to lots of jobs and lots of companies through job boards, which make it easy to spam employers with applications. Our hiring rates from job boards were so low that in 2012 we stopped posting on them entirely.

Sometimes, though, we just hear about extraordinary people and do whatever it takes to get them, even if it means hiring entire teams and opening up new offices for them. Before becoming VP of People Operations[xx] at the speaker and fitness-band maker Jawbone, Randy Knaflic was a key leader in Google's staffing organization, running technical recruiting for all of Europe, the Middle East, and Africa. He led the recruiting and hiring of our team in Aarhus, Denmark, which would go on to revolutionize the speed at which Web browsers ran. "We knew of this small team of brilliant engineers working from Aarhus," Randy told me. "They sold off their previous company and were trying to figure out what to do next. Microsoft got wind of them and was all over them. Microsoft wanted to hire all of them, but they would have to move to Redmond. The engineers said 'No way.' So we swooped in, ran some aggressive hiring efforts, and said, 'Work from Aarhus, start a new office for Google, build great things.' We hired the entire team and it's this group that built the JavaScript engine in Chrome."

Looking back over the years, Google was fortunate to have founders who from the first day cared about high-quality recruiting. But focusing on quality alone wasn't enough. We also had to cast a wider, more diverse net. And we had to get faster.

The first step to building a recruiting machine is to turn every employee into a recruiter by soliciting referrals. But you need to temper the natural bias we all have toward our friends by having someone objective make the hiring decision. As your organization grows, the second step is to ask your best-networked people to spend even more time sourcing great hires. For some, that may turn into a full-time job.

Finally, be willing to experiment. We learned billboards don't

---

[xx] Randy borrowed the title from us, and I'm delighted that he did.

work because we tried one. Our experience in Aarhus taught us that sometimes it makes more sense to hire a team on their terms than on your own.

But if you already know how to find people, how do you figure out exactly which ones to hire and which to pass on? In the next chapter, I'll explain why it's so hard to make a great hiring decision and tell the story of how a hundred years of science and a few good hunches led to our unique hiring system.

WORK RULES...FOR FINDING EXCEPTIONAL CANDIDATES

☐ Get the best referrals by being excruciatingly specific in describing what you're looking for.
☐ Make recruiting part of everyone's job.
☐ Don't be afraid to try crazy things to get the attention of the best people.

5

Don't Trust Your Gut

Why our instincts keep us from being good
interviewers, and what you can do to hire better

Y ou never get a second chance to make a first impression"
was the tagline for a Head & Shoulders shampoo ad campaign in the 1980s. It unfortunately encapsulates how most
interviews work. There have been volumes written about how "the
first five minutes" of an interview are what really matter, describing
how interviewers make initial assessments and spend the rest of the
interview working to confirm those assessments.[80] If they like you,
they look for reasons to like you more. If they don't like your handshake or the awkward introduction, then the interview is essentially
over because they spend the rest of the meeting looking for reasons
to reject you. These small moments of observation that are then
used to make bigger decisions are called "thin slices."

Tricia Prickett and Neha Gada-Jain, two psychology students
at the University of Toledo, collaborated with their professor Frank
Bernieri and reported in a 2000 study that judgments made in the
first ten seconds of an interview could predict the outcome of the
interview.[81] They discovered this by videotaping real interviews and
then having study participants watch short clips:

> Slices were extracted from each interview, beginning
> with the interviewee knocking on the door and ending

10 seconds after the interviewee took a seat, and shown to naïve observers. Observers provided ratings of employability, competence, intelligence, ambition, trustworthiness, confidence, nervousness, warmth, politeness, likability, and expressiveness. For 9 of the 11 variables, thin-slice judgments correlated significantly with the final evaluation of the actual interviewers. Thus, immediate impressions based on a handshake and brief introduction predicted the outcome of a structured employment interview.

The problem is, these predictions from the first ten seconds are useless.

They create a situation where an interview is spent trying to confirm what we think of someone, rather than truly assessing them. Psychologists call this confirmation bias, "the tendency to search for, interpret, or prioritize information in a way that confirms one's beliefs or hypotheses."[82] Based on the slightest interaction, we make a snap, unconscious judgment heavily influenced by our existing biases and beliefs. Without realizing it, we then shift from assessing a candidate to hunting for evidence that confirms our initial impression.[xxi] Malcolm Gladwell spoke with Richard Nisbett, a psychologist at the University of Michigan, about our unwitting self-deception:

> The basis of the illusion is that we are somehow confident
> that we are getting what is there, that we are able to read off
> a person's disposition.... When you have an interview with
> someone and have an hour with them, you don't conceptu-

[xxi] Confirmation bias is just one of the many ways our unconscious minds cause us—unwittingly—to make bad decisions. In an effort to create a less biased and more inclusive workplace, we've been striving to reduce unconscious bias at Google. We described some of our efforts in the article "You Don't Know What You Don't Know: How Our Unconscious Minds Undermine the Workplace" (*Google* [official blog], September 25, 2014, http://goo.gl/kxxgLz).

alize that as taking a sample of a person's behavior, let alone
a possibly biased sample, which is what it is. What you think
is that you are seeing a hologram, a small and fuzzy image
but still the whole person.[83]

In other words, most interviews are a waste of time because 99.4
percent of the time is spent trying to confirm whatever impression the interviewer formed in the first ten seconds. "Tell me about
yourself." "What is your greatest weakness?" "What is your greatest strength?" Worthless.

Equally worthless are the case interviews and brainteasers used
by many firms. These include problems such as: "Your client is a
paper manufacturer that is considering building a second plant.
Should they?" or "Estimate how many gas stations there are in
Manhattan." Or, most annoyingly, "How many golf balls would fit
inside a 747?" and "If I shrank you to the size of a nickel and put you
in a blender, how would you escape?"

Performance on these kinds of questions is at best a discrete
skill that can be improved through practice, eliminating their utility for assessing candidates. At worst, they rely on some trivial bit
of information or insight that is withheld from the candidate, and
serve primarily to make the interviewer feel clever and self-satisfied.
They have little if any ability to predict how candidates will perform in a job.[84] This is in part because of the irrelevance of the
task (how many times in your day job do you have to estimate how
many gas stations there are?), in part because there's no correlation
between fluid intelligence (which is predictive of job performance)
and insight problems like brainteasers, and in part because there is
no way to distinguish between someone who is innately bright and
someone who has just practiced this skill.

Full disclosure: Some of these interview questions have been
and I'm sure continue to be used at Google. Sorry about that.
We do everything we can to discourage this, as it's really just a

waste of everyone's time. And when our senior leaders—myself included—review applicants each week, we ignore the answers to these questions. As we saw with our billboard, some attempts at assessment just don't work. Happily, the 2013 movie *The Internship*, about two washed-up watch salesmen who decide to become interns at Google, gave the answer to the blender question, so at least that one can't be asked as an interview question anymore.[xxii]

A century of science points the way to an answer

In 1998, Frank Schmidt and John Hunter published a meta-analysis of eighty-five years of research on how well assessments predict performance.[85] They looked at nineteen different assessment techniques and found that typical, unstructured job interviews were pretty bad at predicting how someone would perform once hired. Unstructured interviews have an r^2 of 0.14, meaning that they can explain only 14 percent of an employee's performance.[xxiii] This is

---

[xxii] The "correct" answer is that since the question shrinks you—changing your mass—but doesn't change anything else, your strength-to-mass ratio increases and you could simply jump out of the blender. Vince Vaughn's and Owen Wilson's characters also guessed that the blender would break down (which they knew because they had once sold blenders) and they'd be safe. And then? "You've got two nickel-sized men free in the world," they cried out. "Think of the possibilities!... Sunglass repair? We'd be hell on those little screws! Maybe stick us in those little submarines that they put in people's bodies to fight diseases?...I thought we were stuck in a blender. Now we're saving lives?!?...What a journey!"

[xxiii] I'm simplifying here. To be more precise, r^2 is a measure of how well one or more variables predict an outcome. If the r^2 value is statistically significant and near 100 percent (which rarely happens in social science given the messiness of life!), we can confidently predict our outcome based on the other data in the model. If the r^2 value is closer to 0 percent, predictions will be less accurate. The r^2 values are based on the underlying correlation between variables, or the rate at which events occur together. Neither r^2 nor correlation are measures of causality.

In other words, a high positive correlation (e.g., $r^2 = 0.9$) doesn't mean that A causes B, just that A and B have occurred together. For example, if I go for a run at 6:00 a.m. every morning (if only I had that kind of discipline!) and I let my dog into the yard before I go, the timing of my runs is correlated with the dog going outside (and vice versa) since they tend to happen at the same time, but neither one causes the other. However, if you have a big enough set of data, control for other factors, and do some statistical testing to make sure your results are robust, then correlation

somewhat ahead of reference checks (explaining 7 percent of performance), ahead of the number of years of work experience (3 percent), and well ahead of "graphology," or handwriting analysis (0.04 percent), which I'm stunned that anyone actually uses. Maybe some hospitals test the legibility of doctors' handwriting....

The best predictor of how someone will perform in a job is a work sample test (29 percent). This entails giving candidates a sample piece of work, similar to that which they would do in the job, and assessing their performance at it. Even this can't predict performance perfectly, since actual performance also depends on other skills, such as how well you collaborate with others, adapt to uncertainty, and learn. And worse, many jobs don't have nice, neat pieces of work that you can hand to a candidate. You can (and should) offer a work sample test to someone applying to work in a call center or to do very task-oriented work, but for many jobs there are too many variables involved day-to-day to allow the construction of a representative work sample.

All our technical hires, whether in engineering or product management, go through a work sample test of sorts, where they are asked to solve engineering problems during the interview. According to Urs Hölzle, "We do our interviewing based on really testing your skills. Like, write some code, explain this thing, right? Not look at your resume, but really see what you can do." Eric Veach adds, "The interviews are done [by] a large swath of engineers who ask a lot of very data-oriented kinds of questions. They're not just, you know, 'Tell me about a time when...' They're more like, 'Write me an algorithm to do this.'"

The second-best predictors of performance are tests of general cognitive ability (26 percent). In contrast to case interviews and

---

is a pretty good starting point for making judgments about what works and what doesn't. In the case of hiring, interview performance would of course never cause subsequent job performance, but—after controlling for other variables—it can help predict how well someone will do in the job.

brainteasers, these are actual tests with defined right and wrong answers, similar to what you might find on an IQ test. They are predictive because general cognitive ability includes the capacity to learn, and the combination of raw intelligence and learning ability will make most people successful in most jobs. The problem, however, is that most standardized tests of this type discriminate against non-white, non-male test takers (at least in the United States). The SAT consistently underpredicts how women and non-whites will perform in college. Phyllis Rosser's 1989 study of the SAT compared high school girls and boys of comparable ability and college performance, and found that the girls scored lower on the SAT than the boys.[86] Reasons why include the test format (there is no gender gap on Advanced Placement tests, which use short answers and essays instead of multiple choice); test scoring (boys are more likely to guess after eliminating one possible answer, which improves their scores); and even the content of questions ("females did better on SAT questions about relationships, aesthetics, and the humanities, while males did better on questions about sports, the physical sciences, and business").[xxiv] These kinds of studies have been repeated multiple times, and while standardized tests of this sort have gotten better, they are still not great.[xxv]

[xxiv] Unfortunately, Rosser's study identified this difference but didn't explain why it exists. One emerging possible explanation is that girls and boys are equally well positioned to answer all questions, but each gender falls victim to "stereotype threat," a psychological phenomenon where people tend to perform in line with stereotypes when those stereotypes are made salient to them. For example, research has shown that when test takers are made aware of certain stereotypes just before taking a test, their performance changes. In a foundational study, one group of girls was told before a math test that the test produced gender differences, and gender differences were indeed seen, with girls scoring more poorly than boys. When a different group of girls was told that the same test did not produce gender differences, the differences were not seen in performance. [Source: Stephen J. Spencer, Claude M. Steele, and Diane M. Quinn, "Stereotype Threat and Women's Math Performance," *Journal of Experimental Social Psychology* 35, no. 1 (1999): 4–28.][87]

[xxv] In 2014 the College Board, which designs and administers the SAT, announced they were again revamping the SAT to address these and other issues. Even if this effort is successful, it won't help those of us already in college, applying to graduate

As a proof point, Pitzer College, a liberal arts college in Southern California, made reporting test scores optional for applicants who had at least a 3.5 grade point average (GPA) or were in the top 10 percent of their high school classes. Since then, their average admitted-student GPA has grown 8 percent and they've had a 58 percent increase in students of color.[88]

Tied with tests of general cognitive ability are structured interviews (26 percent), where candidates are asked a consistent set of questions with clear criteria to assess the quality of responses. Structured interviews are used all the time in survey research. The idea is that any variation in candidate assessment is a result of the candidate's performance, not because an interviewer has higher or lower standards, or asks easier or harder questions.

There are two kinds of structured interviews: behavioral and situational. Behavioral interviews ask candidates to describe prior achievements and match those to what is required in the current job (i.e., "Tell me about a time...?"). Situational interviews present a job-related hypothetical situation (i.e., "What would you do if...?"). A diligent interviewer will probe deeply to assess the veracity and thought process behind the stories told by the candidate.

Structured interviews are predictive even for jobs that are themselves unstructured. We've also found that they cause both candidates and interviewers to have a better experience and are perceived to be most fair.[89] So why don't more companies use them? Well, they are hard to develop: You have to write them, test them, and make sure interviewers stick to them. And then you have to continuously refresh them so candidates don't compare notes and come prepared with all the answers. It's a lot of work, but the alternative is to waste everyone's time with a typical interview that is either highly subjective, or discriminatory, or both.

---

school, or out in the workforce, who are unlikely to retake the SAT. [Source: Todd Balf, "The Story Behind the SAT Overhaul," *New York Times Magazine*, March 6, 2014.]

There is a better way. Research shows that combinations of assessment techniques are better than any single technique. For example, a test of general cognitive ability (predicts 26 percent of performance), when combined with an assessment of conscientiousness (10 percent), is better able to predict who will be successful in a job (36 percent). My experience is that people who score high on conscientiousness "work to completion"—meaning they don't stop until a job is done rather than quitting at good enough—and are more likely to feel responsibility for their teams and the environment around them. In other words, they are more likely to act like owners rather than employees. I remember being floored when Josh O'Brien, a member of our tech-support team, was helping me with an IT issue in my first month or so. It was a Friday, and when five o'clock rolled around I told him we could finish on Monday. "That's okay. We work to completion," he told me, and kept at it until my problem was resolved.[90]

So what assessment techniques do we use?

The goal of our interview process is to predict how candidates will perform once they join the team. We achieve that goal by doing what the science says: combining behavioral and situational structured interviews with assessments of cognitive ability, conscientiousness, and leadership.[xxvi]

To help interviewers, we've developed an internal tool called qDroid, where an interviewer picks the job they are screening for, checks the attributes they want to test, and is emailed an interview guide with questions designed to predict performance for that job. This makes it easy for interviewers to find and ask great interview questions. Interviewers can also share the document with others on

---

[xxvi] Melissa Harrell, a PhD member of our People Analytics team, adds, "Switching to structural interviews was a clear choice because they are much more predictive of future performance. Plus the approach is better for diversity because having planned questions and scoring rubrics mitigates our reliance on unconscious biases."[91] Read more about our unconscious bias work at http://goo.gl/UtCBSi.

Sample qDroid screen. © Google, Inc.

the interview panel so everyone can collaborate to assess the candidate from all perspectives.

The neat trick here is that, while interviewers can certainly make up their own questions if they wish, by making it easier to rely on the prevalidated ones, we're giving a little nudge toward better, more reliable interviewing.

Examples of interview questions include:

- Tell me about a time your behavior had a positive impact on your team. (Follow-ups: What was your primary goal and why? How did your teammates respond? Moving forward, what's your plan?)

- Tell me about a time when you effectively managed your team to achieve a goal. What did your approach look like? (Follow-ups: What were your targets and how did you meet them as an individual and as a team? How did you adapt your leadership approach to different individuals? What was the key takeaway from this specific situation?)

- Tell me about a time you had difficulty working with someone (can be a coworker, classmate, client). What made this person difficult to work with for you? (Follow-ups: What steps did you take to resolve the problem? What was the outcome? What could you have done differently?)

One early reader of this book, when it was still a rough draft, told me, "These questions are so generic it's a little disappointing." He was right, and wrong. Yes, these questions are bland; it's the answers that are compelling. But the questions give you a consistent, reliable basis for sifting the superb candidates from the merely great, because superb candidates will have much, much better examples and reasons for making the choices they did. You'll see a clear line between the great and the average.

Sure, it can be fun to ask "What song best describes your work ethic?" or "What do you think about when you're alone in your car?"—both real interview questions from other companies—but the point is to identify the best person for the job, not to indulge yourself by asking questions that trigger your biases ("OMG! I think about the same things in the car!") and don't have a proven link to getting the job done.

We then score the interview with a consistent rubric.[xxvii] Our own version of the scoring for general cognitive ability has five constituent components, starting with how well the candidate understands the problem.

For each component, the interviewer has to indicate how the candidate did, and each performance level is clearly defined. The interviewer then has to write exactly how the candidate demonstrated their general cognitive ability, so later reviewers can make their own assessment.

[xxvii] Also known as "behaviorally anchored rating scales."

Upon hearing about our interview questions and scoring sheets, the same skeptical friend blurted, "Bah! Just more platitudes and corporate speak." But think about the last five people you interviewed for a similar job. Did you give them similar questions or did each person get different questions? Did you cover everything you needed to with each of them, or did you run out of time? Did you hold them to exactly the same standard, or were you tougher on one because you were tired, cranky, and having a bad day? Did you write up detailed notes so that other interviewers could benefit from your insights?

A concise hiring rubric addresses all these issues because it distills messy, vague, and complicated work situations down to measurable, comparable results. For example, imagine you're interviewing someone for a tech-support job. A solid answer for "identifies solutions" would be, "I fixed the laptop battery like my customer asked." An outstanding answer would be, "I figured that since he had complained about battery life in the past and was about to go on a trip, I'd also get a spare battery in case he needed it." Applying a boring-seeming rubric is the key to quantifying and taming the mess.

If you don't want to build all this yourself, it's easy enough to find online examples of structured interview questions that you can adapt and use in your environments. For example, the US Department of Veterans Affairs has a site with almost a hundred sample questions at www.va.gov/pbi/questions.asp. Use them. You'll do better at hiring immediately.

Remember too that you don't just want to assess the candidate. You want them to fall in love with you. Really. You want them to have a great experience, have their concerns addressed, and come away feeling like they just had the best day of their lives. Interviews are awkward because you're having an intimate conversation with someone you just met, and the candidate is in a very vulnerable position. It's always worth investing time to make sure they feel good at the end of it, because they will tell other people about their experience—and because it's the right way to treat people.

Sometimes this is as simple as leaving time for conversation. It's too easy in an interview to focus on your needs: You're busy and need to assess this person as fast as you can. But they're making a bigger decision than you are. After all, companies have many employees, but a person has only one job. I make a point of always asking candidates how the recruiting process has been so far, and leaving at least ten minutes for their questions.

Following the interviews, we survey every interviewee with a tool we call VoxPop,[xxviii] to find out what they thought of the process, and later use their feedback to adjust our process accordingly. Based on VoxPop, we now try to build in a quick tour around the office, offer lunch if time permits, and require that every interviewer leave five minutes for the candidate to ask questions. Candidates also told us that we took too long to reimburse them for travel, so we cut that time by more than half.

In contrast to the days when everyone in Silicon Valley seemed to have a story about their miserable Google experience, today 80 percent of people who have been interviewed and *rejected* report that they would recommend that a friend apply to Google. This is pretty remarkable considering that they themselves didn't get hired.

Now you know how to ask interview questions. How do you pick which ones to ask?

We used to think it was sufficient to hire the smartest people we could. But a team of Garry Kasparovs might not be best suited to

---

[xxviii] VoxPop is short for *vox populi*, Latin for "voice of the people."

working together to solve really big problems. So in 2007 we started looking for themes across the ten thousand or so people we had hired, as well as the millions we didn't. In addition to testing technical hires on their engineering ability, we realized that there were four distinct attributes that predicted whether someone would be successful at Google:

1. General Cognitive Ability. Not surprisingly, we want smart people who can learn and adapt to new situations. Remember that this is about understanding how candidates have solved hard problems in real life and how they learn, not checking GPAs and SATs.

2. Leadership. Also not surprising, right? Every company wants leaders. But Google looks for a particular type of leadership, called "emergent leadership." This is a form of leadership that ignores formal designations—at Google there is rarely a formal leader of any effort. I recall being asked what it meant that I was the "executive sponsor" for a project that culminated in 10 percent salary increases for everyone in the company. I explained that not only did I not know, but that within Google it was a meaningless designation. In all likelihood, a relatively new hire had put those words next to my name because my title was Senior Vice President, but my role on the project was the same as anyone else's: Provide an opinion, do some analysis, and help get the right outcome. At Google we expect that over a team's life, different skills will be needed at different times, so various people will need to step into leadership roles, contribute, and—just as important—recede back into the team once the need for their specific skills has passed. We have a strong bias against leaders who champion themselves: people who use "I" far more

than "we" and focus exclusively on what they accomplished, rather than how.

3. "Googleyness." We want people who will thrive at Google. This isn't a neatly defined box, but includes attributes like enjoying fun (who doesn't?), a certain dose of intellectual humility (it's hard to learn if you can't admit that you might be wrong), a strong measure of conscientiousness (we want owners, not employees), comfort with ambiguity (we don't know how our business will evolve, and navigating Google internally requires dealing with a lot of ambiguity), and evidence that you've taken some courageous or interesting paths in your life.

4. Role-Related Knowledge. By far the least important attribute we screen for is whether someone actually knows anything about the job they are taking on. Our reasoning and experience is that someone who has done the same task—successfully—for many years is likely to see a situation at Google and replicate the same solution that has worked for them. As the psychologist Abraham Maslow wrote: "I suppose it is tempting, if the only tool you have is a hammer, to treat everything as if it were a nail."[92] The problem with this approach is that you lose the opportunity to create something new. In contrast, our experience is that curious people who are open to learning will figure out the right answers in almost all cases, and have a much greater chance of creating a truly novel solution.[xxix] For technical roles, such as those in engineering or product management, we assess expertise

---

[xxix] Obviously, there are some roles where specific content expertise is required. You don't want your Tax Department staffed entirely with people who don't know how to fill out a tax return. But even in those departments, we try to mix in people with different backgrounds and novel thinking.

in computer science quite extensively, but even there our bias is to hire people with a general (though expert-level) understanding of computer science rather than specialized knowledge of just one field. And to be fair, we have moved from a philosophy of hiring exclusively generalists to a more refined approach, where we look across our portfolio of talent and ensure we have the right balance of generalists and experts. One of the luxuries of scale is that you can build areas of deep specialization, but even in those pockets we monitor to make sure there is always an influx of fresh, non-expert thinking.

Once we identified these attributes, we began requiring all interview feedback to comment specifically on each one. Not every interviewer had to assess *every* attribute, but at least two independent interviewers had to assess *each* attribute. In addition, we required that the written feedback include the attribute being assessed, the question asked, the candidate's answer, and the interviewer's assessment of that answer. This format would prove to be very valuable because it allowed subsequent reviewers of each candidate to independently assess the candidate. In other words, if you interviewed me and weren't impressed, but had written down your question and my answer, a later reviewer could make their own assessment of whether my responses were good or not. (Of course, getting this level of detail can be awkward—almost every interview at Google starts with the interviewer asking, "Do you mind if I take notes?" Some interviewers even take notes on their laptops, which can be a little disconcerting for the candidate.) Not only does that allow us to give candidates a second chance of sorts, but it also helps us assess whether the interviewer herself is good at assessing people. If we see a consistent pattern of an interviewer "getting it wrong," we either train her or ask her to stop interviewing.

Constantly check that your hiring process actually works

As you can tell, we invest a lot in hiring great people. But our operating assumption is that anything we're doing, we can do better. The first Google search index in 1998 had twenty-six million unique Web pages. By 2000, it had one billion. By 2008, it contained one trillion (1,000,000,000,000!).

According to Jesse Alpert and Nissan Hajaj from our search team, we've made our search engine more comprehensive and efficient: "Our systems have come a long way since the first set of Web data Google processed to answer queries. Back then, we did everything in batches: One workstation could compute the PageRank graph [the algorithm that prioritizes search results] on 26 million pages in a couple of hours, and that set of pages would be used as Google's index for a fixed period of time. Today [in 2008], Google crawls the Web continuously, collecting updated page information and re-processing the entire Web-link graph several times per day. This graph of one trillion URLs is similar to a map made up of one trillion intersections. So multiple times every day, we do the computational equivalent of fully exploring every intersection of every road in the United States. Except it'd be a map about 50,000 times as big as the US, with 50,000 times as many roads and intersections." And, of course, that was more than five years ago. Google Now, introduced in 2012, anticipates what you need to know. For example, your phone can get your boarding pass for an upcoming flight, notify you that traffic is heavy on the freeway so you can take a side street, or tell you about cool events nearby.

And just as our products can always get better, so can our hiring machine. We constantly review and work to balance our speed, error rate, and quality of experience for candidates and Googlers. For example, Todd Carlisle, now the HR leader for one of our business teams but at the time a PhD analyst on our staffing team, looked at the question of whether having up to twenty-five interviews per candidate was actually helpful or not. He found that four interviews were enough to predict whether or not we should hire someone with 86 percent confidence. Every additional interviewer after the fourth added only 1 percent more predictive power. It simply wasn't worth the extra time for Google or the suffering for the candidate, so we implemented a "Rule of Four," limiting the number of interviews a candidate could have on-site (though we allowed exceptions in certain cases). That change alone shaved our median time to hire to 47 days, compared to 90 to 180 days in the past, and has saved employees hundreds of thousands of hours.

To this day, we never assume we get it right each time. We revisit the applications of rejected candidates to assess if we made errors, and to correct them and learn from them. Our Revisit Program starts by feeding the resumes of all current incumbents for a particular job, such as software engineer, into an algorithm that identifies the most common keywords. That list is then reviewed and augmented by a handpicked group of recruiters and managers. For example, if IEEE (the Institute of Electrical and Electronics Engineers) turns up as a common keyword, they might add the names of other professional associations. The updated list of keywords is run through another algorithm, this time looking at the past six months of applicants and assigning a weighting for each keyword depending on how frequently it occurs on successful and unsuccessful resumes. Finally, we score the next six months of incoming resumes against these weighted keywords and flag candidates who were both rejected and had high scores, so that our recruiters can take a second look. In 2010, we ran 300,000 rejected software engineer resumes through this system, revisited 10,000 applications, and hired 150 people. This may seem like a lot of work to get 150 hires, but a yield of 1.5 percent is six times better than our overall hiring yield of 0.25 percent.

We don't just look at the candidate side of hiring. Interviewers also receive feedback on their own personal ability to predict whether someone should be hired. Every interviewer sees a record of the interview scores they have given in the past and whether those people were hired or not.

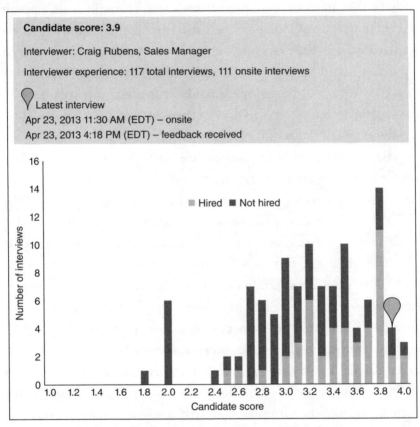

Illustrative feedback for an interviewer. © Google, Inc.

This lets the interviewer know if they are correctly assessing potential Googlers, nudging them to look back at their prior interview notes and learn from what they spotted or missed. And it lets later reviewers of each candidate's packet know whether any given interviewer is reliable or should be ignored.

Never compromise on quality

We've focused so far on finding candidates and interviewing, but these are just two parts of the hiring process. Superficially, every organization's hiring process looks the same, and it's pretty boring stuff. Post a job. Get resumes. Review them. Interview people. Hire one. Zzzzz…

Dig more deeply, and the Google approach starts looking very different as soon as a candidate applies for a job. There are six unique parts of our screening process, with a goal of ensuring that the bar for quality is never compromised and that our decisions are as free of bias as possible.

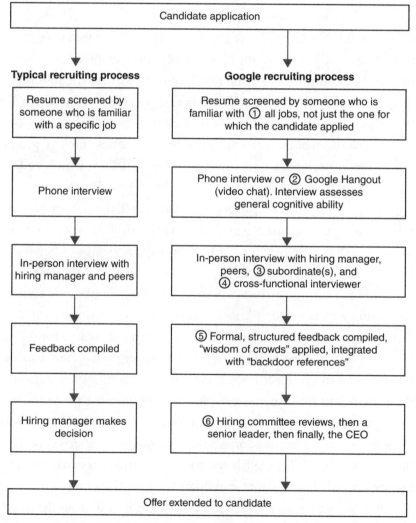

Comparison of a typical hiring process with Google's hiring process. © Google, Inc.

First, assessment is led by Google's dedicated recruiters rather than line managers. Our recruiters are experts in interpreting resumes, which is critical when you're receiving resumes from more than a hundred countries. For example, when evaluating college students, GPA might seem like a fairly important factor to consider. Not so for candidates from Japan. In Japan, college admissions are largely based on national test results, so high school students focus intensely on doing well on those tests, often attending *juku* (special after-school classes) for fifteen to twenty hours each week for years. But once admitted to a premier university, Japanese students don't focus on grades at all. Historically, they enjoy a last gasp of play and freedom between the crush of the *juku* and the monotony of the career of a *sarariman* ("salaryman"—the nomenclature for the expected rule-following, slow, tenure-based progression that characterized Japanese careers in the past). Japanese college grades are virtually useless as a hiring signal, but knowing which college someone attended is helpful, at least for hiring new graduates.

Our professional recruiters are also familiar with many jobs across Google, no small feat considering our business currently includes search, self-driving cars, futuristic glasses, fiber-based Internet services, manufacturing, video studios, and venture capital! This is important, because when someone applies to a job at your company, they don't know everything your company does. In fact, most large companies have distinct recruiting teams for different divisions. Someone rejected for a product management job in one division might have been great for a marketing job in another division, but won't be considered for that job because the recruiters in the two divisions don't talk. At Google, a person rejected from a job as product manager for Android might be a superb candidate for a sales role working with telecommunications companies. Our recruiters are able to route candidates across the entire company, which requires both visibility into all the jobs and an understanding of what they are. And if there's no job available at the moment, the recruiters make a note to follow up with strong candidates for future opportunities.

After the resume is screened and selected, the second part of our process is a remote interview. These are much harder to conduct than in-person interviews because it's difficult to build rapport and pick up on nonverbal cues. Phone interviews are particularly challenging for people who are not fluent in English (the language of the company is English), as it's harder to make yourself understood over a phone. We prefer to use Google Hangouts, which provide video interaction and allow screen and whiteboard sharing so that technical candidates and interviewers can write and review software code together. Hangouts require no special equipment, conference centers, or downloads. Candidates simply log on to Google+ and receive a pop-up inviting them to join a Hangout: instant video conference. Using Hangouts also minimizes costs, as a remote interview is far less expensive than an in-person one, and is more respectful of Googlers' and candidates' time. Our recruiters have the benefit of having done them hundreds of times, compared to a typical hiring manager who may have done remote assessments only once or twice.

Having professionals do the initial remote evaluation also means that it's possible to do a robust, reliable screening for the most important hiring attributes up front. Often a candidate's problem-solving and learning ability is assessed at this stage. We do this early so that later interviewers can focus on other attributes, like leadership and comfort with ambiguity.

As an aside, professional recruiters are also a little more ready to roll with the punches when something nutty happens in an interview. Like the candidate who brought his mom. Or the engineering candidate who forgot to wear a belt, so his pants started falling down every time he turned to write code on the whiteboard. Our seasoned recruiter came to the rescue and gave the candidate his own belt.

In every interview I've ever had with another company, I've met my potential boss and several peers. But rarely have I met anyone who would be working for me. Google turns this approach upside down. You'll probably meet your prospective manager (where possible—for some large job groups like "software engineer" or "account strategist" there is no single hiring manager) and a peer, but more important is meeting one or two of the people who will work for you. In a way, their assessments are more important than anyone else's—after all, they're going to have to live with you. The third key difference in our approach, therefore, is to have a subordinate interview a prospective hire. It sends a strong signal to candidates about Google being nonhierarchical, and it also helps prevent cronyism, where managers hire their old buddies for their new teams. We find that the best candidates leave subordinates feeling inspired or excited to learn from them.

Fourth, we add a "cross-functional interviewer," someone with little or no connection at all to the group for which the candidate is interviewing. For example, we might ask someone from the legal or the Ads team (the latter design the technology behind our advertising products) to interview a prospective sales hire. This is to provide a disinterested assessment: A Googler from a different function is unlikely to have any interest in a particular job being filled but has a strong interest in keeping the quality of hiring high. They are also less susceptible to the thin-slices error, since they have less in common with the candidate than the other interviewers.

Fifth, we compile feedback about candidates in a radically unusual way. We've discussed how interview feedback must cover our hiring attributes, and the use of backdoor references. In addition, we equally weight each individual's feedback on the candidate. A subordinate's feedback is at least as valuable, if not more so, as a hiring manager's. Todd's research showed not only that the optimal number of interviews was four, but that virtually no single interviewer's assessment was by itself that helpful.

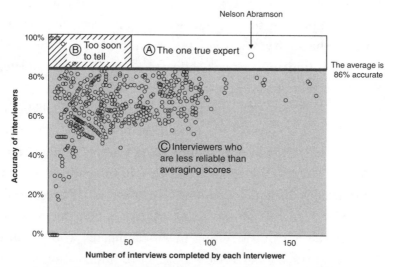

Scatterplot showing individual interviewing accuracy (each dot is one interviewer) compared to the 86 percent accuracy resulting from averaging interview scores. Accuracy of interviewers is defined as the percent of candidates that the interviewer wants to hire who are actually hired. Group A consists of Nelson Abramson, the only person to beat the wisdom of the crowd. In Group B, these interviewers seem more accurate than the crowd but haven't done enough interviews for us to statistically prove if it's skill or if they were just lucky. Most interviewers are less accurate than "the crowd" and fall into Group C.

Or, to be fair, only one Googler's was: Nelson Abramson. He's the lone dot on the far top right of the graph. But when Todd dug into it, he discovered Nelson had an unfair advantage. He worked in our data centers, a global network of servers that effectively make copies of the Internet so that search results can come to you in milliseconds. This role requires a very distinctive skill set, and he interviewed people only for these jobs. He was also employee number 580, so he'd enjoyed lots of practice. But that was the only such case we found across the five thousand on-site engineering interviews in the analysis.

As has been found in other settings,[xxx] the "wisdom of the crowds" seemed to apply to making hiring decisions as well.

So we continue reporting the individual interview feedback scores but emphasize the average score.[93] This has the virtue of eliminating the ability of any single person to blackball any candidate, as well as limiting anyone's ability to politick for a candidate.

Sixth, we rely on disinterested reviewers. In addition to using structured interviews and the hiring attributes, we deliberately include at least three layers of review for each candidate. The hiring committee takes a first look, recommending whether or not to move a candidate forward. For example, in People Operations, the hiring committee is made up of various directors and VPs responsible for the major segments on our team. Hiring-committee members recuse themselves from weighing in on candidates who would be on their own teams. They review a packet of information for each candidate that is forty to sixty pages long. I've excerpted some of the key elements on the following page.

If the hiring committee rejects the candidate, the process stops there. If they are supportive of a candidate, their feedback is added to the hiring packet and sent to the Senior Leader Reviews. In these weekly meetings, some of our top executives provide another layer of objectivity and assess that week's candidates. We've had weeks with more than three hundred candidates and weeks with only twenty. At this stage, candidates are recommended for hire, rejected, or more information is requested, typically to test an attribute further or to reconsider the level at which the candidate would be hired. The most common reason for rejection at this stage in the

---

[xxx] For example, Scott Page of the University of Michigan has shown that averaging the guesses of President Obama's finance team about what financial markets will do is more accurate than the analysis put out by a small group of economists at the Federal Reserve. On *Who Wants to Be a Millionaire*, asking the audience gets you the right answer 95 percent of the time, according to former host Regis Philbin. And Google's PageRank, an algorithm that prioritizes search results, relies heavily on the wisdom of crowds.

Candidate's Name

EXTERNAL REFERENCE CHECKS

Name of Reference

Company Name: Fortress Investment Group   Phone Number: 415-284-7423

Check Comp

1. W

Role-related   **INTERVIEWER SCORE AND FEEDBACK**
FOR EACH INTERVIEW QUESTION

Leveraging on
consumers jo
knowledge and insights for our clients is crucial for the role she is applying for and Google's

GCA   ends in the large
DETAILED FEEDBACK ON EACH CANDIDATE   stment companies
Unde   y create
Identi   **ANSWER, BASED ON FOUR HIRING ATTRIBUTES**   ss is a crucial
Makes Sound Decisions: Solid   the objective of
                                                                       brings to the role.

Offer Coversh   **DETAILED CANDIDATE BACKGROUND**   Offer ID:
Position/Title: Child   74396

Recruiter: Lead Recruiter:

Candidate Summary

Type of Hire: Direct Hire

Education

School, Country	Selectivity	Degree, Field			Grad Date	Graduated?
San Jose State University		Undeclared			December 2003	NO
Ohlone College		Associate Degree, Early Childhood Studies			May 2004	NO
California State University, Sacramento		Bachelor Arts, Child Development			May 2006	YES

Work History

Company/Organization	Position	Start Date	Term Date
Children's Creative Learning Centers, Inc. - Cisco Campus	Pre-Kindergarten Math and Phonics Instructor, Pre-Kindergarten Head Teacher, Preschool Head Teacher	November 2008	In Progress
Children's House of Los Altos	Preschool Teacher/School Age Program Director	August 2006	June 2008

Years of Relevant Industry Experience: 4

Interview Data

Number of Interviews: 5   Average Score: 3.4

Additional Notes

Per   Director of Benefits: I support hiring   as an Infant Toddler Support Teacher at the Children's Center.   received a BA from California State University, Sacramento in Child Development. She currently works at a play based center and is eager to join Google's program which is inspired by the principles of the Reggio Emilia approach and fosters collaborative learning and developmental growth. She has several years of experience working at Children's Creative Learning Centers at the Cisco Campus so she is familiar with the corporate element of working with employee sponsored child care. She also spoke strongly about working collaboratively with teachers, parents and children. Children's Center Director   spoke with   and   thought the candidate was very poised and articulate displaying good values and core thinking that is in line with the Google Children's Centers. For these reasons, I think she would make a good Infant Toddler Support teacher at the Google Children's Centers.

RULE OF 7:
* Will report to: Site Director
* Manager's team size: 26
* People Manager?: No.

GCA INTERVIEWER: Google Children's Center, Operations Manager

Hiring packet excerpts. © Google, Inc.

process? Culture.[xxxi] While Googlers possess the gamut of political views, the cultural values of transparency and voice are widely held and core to how we operate. As Jeff Huber recently said about one candidate, "This is a great candidate—strong technical interview scores, clearly very smart and well-qualified—but sufficiently arrogant that none of the interviewers want him on their team. This is a great candidate, but not for Google."

If the Senior Leader Review is supportive, then Larry is sent each week's recommended hires. The report includes links to the detailed hiring packets for every candidate as well as summaries about each candidate and the feedback and recommendations from each successive level of review. The most common feedback from Larry is that a candidate might not meet our hiring bar or that the creativity shown in a portfolio might not be up to snuff. More important than the feedback itself is the message from Larry to the company that hiring is taken seriously at the highest levels, and that we have a duty to continue doing a good job. And new Googlers ("Nooglers") are always delighted to learn that Larry personally reviewed their applications.

If we followed a more traditional process, it's likely we could hire people in a week or two instead of the six weeks it takes today. And we do have the capability to move more quickly when needed—every few weeks we'll run an expedited process for candidates who have an offer from another firm that expires if they don't respond quickly,[xxxii] and we've run one-day hiring programs

---

[xxxi] Culture here refers specifically to the attributes described earlier in the chapter, including conscientiousness, comfort with ambiguity, etc. It's also about broadening the kinds of people who work at Google and avoiding homogeneity.

[xxxii] These are called "exploding offers" because the offer goes away (explodes) if not accepted by a certain date. They are common in college hiring, but are showing up more and more in Silicon Valley. I think they put a lot of unfair pressure on the candidate, who should be free to make the best decision for herself without duress. After all, companies have lots of employees, but each person holds only one job. It should be one they are sure of.

on college campuses in the United States and India to test if that improved our offer acceptance rates. So far, the greater speed doesn't materially improve candidate experience or the rate at which candidates accept our offers, so our focus remains on finding ways to hire people we might overlook rather than on moving faster.

Putting it all together: how to hire the best

If you wondered if this takes a lot of Googler time, it does.

But not nearly as much as you might think. There are four simple principles that can help even the smallest team do much, much better at hiring.

Until we hit about twenty thousand employees, most people in the company spent four to ten hours per week on hiring, and our top executives would easily spend a full day each week on it, which worked out to between eighty thousand and two hundred thousand hours per year spent on hiring. And this excluded the time spent by our staffing teams. That was necessary to grow quickly and to ensure we didn't ever compromise quality. And, honestly, it was the best we could do at the time. It took years of research and experimentation to figure out how to hire more efficiently.

In 2013, with roughly forty thousand people, the average Googler spent one and a half hours per week on hiring, even though our volume

---

Robert J. Robinson, at the time a Harvard professor, described how to counter an exploding offer in his 1995 piece "Defusing the Exploding Offer: The Farpoint Gambit" (*Negotiation Journal* 11, no. 3: 277–285). The title refers to a *Star Trek: The Next Generation* episode called "Encounter at Farpoint." The captain of the starship *Enterprise* is being tried by an alien judge who tells the guards, "Soldiers, you will press those triggers [on your guns] if this criminal answers with any word other than 'guilty.'" His next question to Captain Picard is, "Criminal, how do you plead?" to which Picard answers, "Guilty... provisionally." Since Picard is now both not guilty ("provisionally") and not dead, the coercive condition falls away.

Robinson's clever idea is to do the same with exploding offers: "I accept, provisionally." The gambit is to present a reasonable condition (e.g., "provided I can meet with the person I will be working for"; "provided I don't get a better offer from the companies I'm still waiting to hear from") that causes the deadline to pass. "Once the deadline passes," Robinson argues, "the credibility of the threat is destroyed."

of hiring is almost twice what it was when we had twenty thousand people. We've reduced the amount of time spent by Googlers on each hire by about 75 percent. We continue to work to reduce this, and to become more efficient in managing our staffing teams and their time.

But by far the best recruiting technique is having a core of remarkable people. Jonathan Rosenberg used to keep a stack of two hundred Googler resumes in his office. If a candidate was on the fence about joining Google, Jonathan would simply give them the stack and say: "You get to work with these people." These Googlers came from every kind of educational background, including many of the best schools on the planet; had invented seminal products and technologies like JavaScript, BigTable, and MapReduce; had been part of some of the most revolutionary companies; and included Olympic athletes, Turing Award and Academy Award winners, Cirque du Soleil performers, cup stackers, Rubik's Cube champions, magicians, triathletes, volunteers, veterans, and people who had accomplished just about any cool thing you could think of. Invariably, the candidate would ask if Jonathan had cherry-picked the resumes. And he would honestly tell them that it was a random sample of the people who build Google's products. He never lost a candidate.

So how do you create your own self-replicating staffing machine?

1. **Set a high bar for quality.** Before you start recruiting, decide what attributes you want and define as a group what great looks like. A good rule of thumb is to hire only people who are better than you. Do not compromise. Ever.

2. **Find your own candidates.** LinkedIn, Google+, alumni databases, and professional associations make it easy.

3. **Assess candidates objectively.** Include subordinates and peers in the interviews, make sure interviewers write good notes, and have an unbiased group of people make the actual hiring decision. Periodically return to those notes and com-

pare them to how the new employee is doing, to refine your
assessment capability.

4. **Give candidates a reason to join.** Make clear why the work
you are doing matters, and let the candidate experience the
astounding people they will get to work with.

This is easy to write, but I can tell you from experience that
it's very hard to do. Managers hate the idea that they can't hire
their own people. Interviewers can't stand being told that they have
to follow a certain format for the interview or for their feedback.
People will disagree with data if it runs counter to their intuition
and argue that the quality bar doesn't need to be so high for every
job.

Do not give in to the pressure.

Fight for quality.

I'm often told: "I just want an administrative assistant who can
answer the phone and schedule meetings—I don't need someone
brilliant, just someone who can do the work." But that's terrible
logic. An outstanding administrative assistant provides powerful
leverage to a manager, helping them better allocate their time, pri-
oritizing and routing less critical tasks, and being the face of the
manager to everyone who reaches out. These roles are important,
and the difference between an average administrative assistant and
an exceptional one is profound. And I should know, as I'm privileged to work with one of the best, Hannah Cha.

If you're committed to transforming your team or your organization, hiring better is the single best way to do it. It takes will and
patience, but it works. Be willing to concentrate your people investment on hiring. And never settle.

There's one other beneficial effect of hiring this way: In most
organizations, you join and then have to prove yourself. At Google,

there's such faith in the quality of the hiring process that people join and on their first day are trusted and full members of their teams.

There's no better illustration of this than when, in 2011, the Dalai Lama was invited by Archbishop Desmond Tutu to deliver the inaugural Desmond Tutu Peace Lecture in Cape Town on the archbishop's eightieth birthday. Both Nobel Peace Prize laureates, this was to be an historic conversation. But after allegedly receiving pressure from the Chinese government, the governing African National Congress decided they would not issue a travel visa for the Dalai Lama.[94] Archbishop Tutu was incensed: "Our government—representing me!—says it will not support Tibetans being viciously oppressed by China. You, President Zuma and your government, do not represent me." Loren Groves, a Noogler who had just launched his first product at Google a week earlier, was dispatched to Tibet and then to South Africa. He arranged a meeting between the two men via Google Hangout, allowing them to have a face-to-face con-

OCTOBER 5, 2011 *8*

NEW DELHI — The Dalai Lama, Tibet's exiled spiritual leader, ~~scrapped plans on Tuesday to attend~~ *joined* the 80th birthday celebration of a fellow Nobel laureate, Desmond M. Tutu of South Africa, *via hangout* after the host government did not grant his visa request.

Advertisement in the *New York Times* celebrating the successful Hangout between the Dalai Lama and Desmond Tutu. © Google, Inc.

versation despite being thousands of miles apart. This was the centerpiece of the celebration on October 8.[95]

The next day we ran a full-page ad in the *New York Times*.

The original, signed by the Dalai Lama and Archbishop Tutu, hangs in our offices. As the conference's moderator explained: "While it is a real pity that His Holiness cannot be here, we are so grateful for the world of technology that has come to us. And that we can have this conversation."

And it was thanks to Loren, who had been a Googler for five days and whom we trusted to make this encounter possible.

- -

WORK RULES...FOR SELECTING NEW EMPLOYEES

☐ Set a high bar for quality.
☐ Find your own candidates.
☐ Assess candidates objectively.
☐ Give candidates a reason to join.

- -

6

·····················

Let the Inmates Run the Asylum

Take power from your managers and trust your
people to run things

Does your manager trust you?
I'm sure she doesn't hide her jewelry when you enter the
room, but if you thought you were ready for a promotion,
could you promote yourself? If you wanted to spend one day a week
working on a side project or organizing lectures for other employees,
and you figured out a way to still get your job done, could you? Is
there a limit to how many sick days you can take?

Just as important, do you trust your manager? Does she sponsor and fight for you and help you get work done? If you're thinking
about taking another job, can you talk to her about it?

This is the kind of manager we'd all love to have, but few of us
have actually enjoyed. At Google, we have always had a deep skepticism about management. This is just how many engineers think:
Managers are a Dilbertian layer that at best protects the people
doing the actual work from the even more poorly informed people
higher up the org chart.

But our Project Oxygen research, which we'll cover in depth
in chapter 8, showed that managers in fact do many good things. It
turns out that we are not skeptical about managers per se. Rather,
we are profoundly suspicious of power, and the way managers historically have abused it.

A traditional manager controls your pay, your promotions, your workload, your coming and going, whether you have a job or not, and these days even reaches into your evenings and weekends. While a manager doesn't necessarily abuse any of these sources of power, the potential for abuse exists. Our anxieties about toxic bosses show up everywhere in the culture, from Michael Scott on *The Office* to the recent flood of books like *The No A**hole Rule* and *A**holes Finish First* (the first teaches how to survive working with jerks and the latter how to be one).[xxxiii]

When I worked at GE, I knew a senior executive I'll call Ellen. Ellen had fast-tracked through GE and been rewarded with a top job. One morning, Ellen breezed into her office and dropped a small paper bag on her secretary's desk. "Lisa, can you run this to my doctor's office? I have to give him a stool sample." The bag contained a still-warm piece of Ellen's morning production.

Ellen didn't see anything wrong with what she had done. She was a busy executive, and having her secretary haul her excrement around town simply made Ellen more efficient.

You may have heard the phrase "Power corrupts; absolute power corrupts absolutely."[96] When Lord Acton wrote those words in 1887, he was making a deeper point about leadership. He was arguing with Mandell Creighton, a historian and bishop of the Church of England, who was writing a history of the Inquisition that somewhat absolved the pope and king of responsibility. Acton made an even more forceful argument than most know:

> I cannot accept your canon that we are to judge Pope and King unlike other men, with a favourable presumption that they did no wrong. If there is any presumption it is the other way against holders of power, increasing as the power increases....Great men are almost always bad men, even

---

[xxxiii] The asterisks are mine. This is a family book, after all.

when they exercise influence and not authority: still more when you superadd the tendency or the certainty of corruption by authority. There is no worse heresy than that the office sanctifies the holder of it. That is the point at which . . . the end learns to justify the means.

Acton isn't just making some academic observation that power corrupts. He's shouting that those in authority must be held to even higher standards than the rest.

Against this backdrop, Ellen's actions are less surprising. After all, hadn't she worked hard and sacrificed to earn that senior executive title? Surely she was so busy that if her secretary could save her even fifteen minutes, it would be worthwhile for GE, as Ellen could channel that precious time into creating more shareholder value? And if it spilled from a professional need to a personal one, well, Ellen frequently did GE's work on her personal time. Helping her personally wasn't that different from helping her professionally, right?

Wrong.

Managers aren't bad people.

But each of us is susceptible to the conveniences and small thrills of power.

At the same time, responsibility for creating (and fighting!) hierarchy doesn't fall solely on the shoulders of managers. We employees often create our own hierarchies.

One of the challenges we face at Google is that we want people to feel, think, and act like owners rather than employees. But human beings are wired to defer to authority, seek hierarchy, and focus on their local interest. Think about meetings that you go to. I'd wager that the most senior person always ends up sitting at the head of the table. Is that because they race from office to office, scurrying to be there first so they can seize the best seat?

Watch closely next time. As attendees file in, they leave the head seat vacant. It illustrates the subtle and insidious nature of how we create hierarchy. Without instruction, discussion, or even conscious thought, we make room for our "superiors."

I see this even at Google, but with a twist. Some of our most senior leaders are attuned to this dynamic, and have tried to break it by sitting at the center of a conference table, along one of the sides. Kent Walker, our general counsel, regularly does so. "In part it's to create a 'King Arthur's Round Table' dynamic—less hierarchical and more calculated to draw people into a conversation with each other rather than a series of back-and-forth exchanges with me."[xxxiv]

Invariably, within a few meetings, that's the seat that ends up left open.

Humans turn out to be awfully good rule followers. Before 2007, the hiring policy at Google was "Hire as many great people as you can." In 2007, we introduced hiring budgets because we were hiring more people than we could absorb. Each team now had a finite number of people they could hire each year. I was stunned by how quickly we shifted from an abundance mentality to one of scarcity, as jobs became a precious resource that had to be conserved. Roles would be held open longer than ever because teams wanted to be sure they were getting the best person. Internal transfers became more difficult because they required an open headcount slot.

It works a bit better now. We addressed some of these challenges by changing the rules so some teams could go over budget if they needed to—for example, if a Googler wanted to shift over from another team. Most leaders keep a reserve budget as well, so that they always have room for an exceptional hire. But what struck me at the time was that even at a company that aspires to give people

---

[xxxiv] Though Kent is also honest enough to admit it's practical too: "In part it's because I put our meeting agenda on the whiteboard and it's just easiest for me to stay focused when I'm right across from it."

so much freedom, the introduction of simple rules caused large changes in behavior.

The best Googlers apply their own judgment and break the rules when it makes sense. To take a trivial example, we limit Googlers to bringing two guests to our cafés per month. If someone occasionally brings both their parents and their kids, that's fine. It's better that they all have a great experience once in a while than that they conform to a rule.

Now, budgets may seem different. The whole point of a budget is that you're supposed to stay within it. But at Google you should always, always make room for a truly exceptional person, even if it puts you over budget. And yet many of us have such a built-in respect for following norms that it feels revolutionary to suggest that.

Stanley Milgram's controversial experiments at Yale in the 1960s made the same point, but more strenuously. Milgram was exploring the question "How could the Holocaust have happened?" How was it possible that millions of people were murdered not in spite of society, but with the passive and active support of it? Are human beings so susceptible to authority that they would commit the most unconscionable acts?

Presented as a memory experiment, subjects were told to administer shocks to a hidden "learner" if the learner failed to remember the words they were taught. For each failure, the subject would be told to flip a switch increasing the voltage by 15 volts, rising from 15 to 420 volts, with two final switches labeled XXX and corresponding to 435 and 450 volts. At each increment, the subject would hear recordings of the learner shouting, then screaming. At 300 volts, the learner would start pounding on the wall and complaining about his heart condition. After 315 volts, the learner would fall silent. The experiment would be halted when the subject refused to flip any more switches, or after he had shocked with 450 volts—in some versions of the experiment up to three times. It took thirty-one shocks to get to that point.

In Milgram's first experiment, forty men participated as subjects. Twenty-six of them went all the way to 450 volts. After the first nineteen shocks, the learner went dead silent. Yet 65 percent of participants kept following orders, administering twelve shocks even after the learner had become completely unresponsive. And of the fourteen who didn't progress all the way, not one asked that the experiment be stopped, or even went to check on the victim without first asking permission.[97]

(A key fact here is that no one was actually electrocuted. The switches didn't do anything, and the screaming was prerecorded.)[xxxv]

Managers have a tendency to amass and exert power.

Employees have a tendency to follow orders.

[xxxv] Milgram's study is often mentioned in passing but the details are revealing. He conducted at least nineteen variations of this experiment. Experiment 8 consisted of only female participants. The obedience results were the same as for men, but women reported higher levels of stress. Regardless of gender, Milgram reported that "Many subjects showed signs of nervousness in the experimental situation, and especially upon administering the more powerful shocks. In a large number of cases the degree of tension reached extremes that are rarely seen in sociopsychological laboratory studies. Subjects were observed to sweat, tremble, stutter, bite their lips, groan, and dig their fingernails into their flesh. These were characteristic rather than exceptional responses to the experiment....Although obedient subjects continued to administer shocks, they often did so under extreme stress. Some expressed reluctance to administer shocks beyond the 300-volt level, and displayed fears similar to those who defied the experimenter; yet they obeyed." [Source: Stanley Milgram, "Behavioral Study of Obedience," *Journal of Abnormal and Social Psychology* 67, no. 4 (1963): 371–378.]

Milgram followed up with study participants to assess whether there were long-term effects. A surprising 84 percent were "very glad" or "glad" about their involvement and 15 percent were neutral. One oft-referenced letter from a participant, quoted by Milgram in *Obedience to Authority* (New York: Harper & Row, 1974), explains that the experiment created a level of self-awareness in participants about how they make decisions that hadn't existed before: "While I was a subject in 1964, though I believed that I was hurting someone, I was totally unaware of why I was doing so. Few people ever realize when they are acting according to their own beliefs and when they are meekly submitting to authority."[98]

What's mind-blowing is that many of us play both roles, manager and employee, at the same time. We each have experienced the frustration of a controlling manager, and we have each experienced the frustration of managing people who just won't listen.

At this point you're probably thinking, Wow, things got pretty dark all of a sudden.

There is hope.

"Does your manager trust you?" is a profound question.

If you believe people are fundamentally good, and if your organization is able to hire well, there is nothing to fear from giving your people freedom.

Remember that the primary definition of "asylum" is "a place of refuge." One of the nobler aspirations of a workplace should be that it's a place of refuge where people are free to create, build, and grow. Why not let the inmates run the asylum?

The first step to mass empowerment is making it safe for people to speak up. In Japan there's a saying: *Deru kugi wa utareru.* "The stake that sticks up gets hammered down." It's a warning to conform.

This is why we take as much power away from managers as we can. The less formal authority they have, the fewer carrots and sticks they have to lord over their teams, and the more latitude the teams have to innovate.

Eliminate status symbols

We discussed how at Google a manager cannot unilaterally hire someone, and in the coming chapters I'll share how we also don't allow managers to make pay and promotion decisions without input from others. But creating an environment of mass empowerment where employees feel and act like owners takes more than managing hiring and promotions in novel ways. To mitigate our innate human tendency to seek hierarchy, we try to remove the signifiers of power and status. For example, as a practical matter there are really only four meaningful, visible levels at Google: individual contributor, manager, director, and vice president. There's also a parallel track for technical people who remain individual contributors throughout their careers. Progression through these levels is a function of a person's scope, impact, and leadership. People of course care about promotions, and promotions to director and executive are very big deals.

When the company was smaller, we drew a public distinction between two levels of director, where the more junior role would be titled as Director, Engineering, and the more senior role would be Engineering Director. We found that even such a subtle distinction as the word order of the title caused our people to fixate on the difference between the levels. So we eliminated the difference.

To be transparent, it's become harder to hold this line as we've gotten bigger. Titles that we used to ban outright, like those containing the words "global" or "strategy," have crept into the company. We had banned "global" because it's both self-evident and self-aggrandizing. Isn't every job global, unless it specifically says it isn't? "Strategy" is similarly grandiose. Sun Tzu was a strategist. Alexander the Great was a strategist. Having been a so-called strategy consultant for many years, I can tell you that putting the word "strategy" in a title is a great way to get people to apply for a job, but it does little to change the nature of the work. We policed the titles as people were hired, but failed to consistently scrub our employee database to catch the titles people gave themselves after joining.[xxxvi] We just hope our efforts make them matter less than in most other places.

Randy Knaflic of Jawbone, whom we met in chapter 4, told me how he's exported the practice of de-emphasizing titles, and that it doesn't work for every person: "At Google, leadership didn't equate to title. I'd often give my top performers leadership opportunities

---

[xxxvi] Yes, you can choose your own title at Google.

and help them learn the art of leading without the titled authority. Over time, it became a no-brainer to move these leaders to positions of people management, as they had thoroughly learned how to inspire leadership, followership, and drive decisions across their peer group. At Jawbone, I tried something similar when I hired an HR business partner from [another technology company]. I explained how title should follow leadership. It was a red flag within the first few weeks when he asked, 'But how can I get them to do what I want them to do if I don't have the title?' He lasted less than six months."

But we also eliminate other signifiers and reinforcers of hierarchy: Our most senior executives receive only the same benefits, perquisites, and resources as our newest hires. There are no executive dining rooms, parking spots, or pensions. When we introduced a deferred compensation program in 2011 (the Google Managed Investment Fund), which allows Googlers to invest their bonus money alongside our finance department, we decided to make it available to everyone rather than just to senior executives, in contrast to what most companies do. In Europe, where it's common for executives to receive car allowances, we offered them to all employees and kept the offering cost neutral by limiting the size of the benefit our more senior people received. Some grumbled, but it was more important to be inclusive than to conform to our industry's practices.

If you want a nonhierarchical environment, you need visible reminders of your values. Otherwise, your human nature inevitably reasserts itself. Symbols and stories matter. Ron Nessen, who served as press secretary for President Gerald Ford, shared a story about his boss's leadership style: "He had a dog, Liberty. Liberty has an accident on the rug in the Oval Office and one of the Navy stewards rushes in to clean it up. Jerry Ford says, 'I'll do that. Get out of the way, I'll do that. No man ought to have to clean up after another man's dog.'"[99]

What makes this vignette so compelling is that the most powerful man in the United States not only understood his personal responsibility, but also appreciated the symbolic value of demonstrating it.

That's why it matters that Patrick Pichette wears jeans and an orange backpack instead of a suit and a briefcase. Yes, he's the CFO of Google, charged with balancing Google's unbounded appetite for moon shots with ensuring our economics are thoughtfully and responsibly managed. But he's also accessible, warm, human. When he rockets around our campus on a bicycle, he's showing that even our most senior leaders are just people.

Patrick (left) and me on one of his bikes.

Make decisions based on data, not based on managers' opinions

In addition to minimizing the trappings and affectations of power, we rely on data to make decisions. Omid Kordestani was at Netscape before coming to Google. As Omid tells it, "Jim Barksdale, the legendary CEO of Netscape, in one of these management meetings said, 'If you have facts, present them and we'll use them. But if you have opinions, we're gonna use mine.'"

The tone of Barksdale's comment is both funny and on the edge of tyrannical, but it captures well how most successful managers think. After all, they (ideally) became managers because they demonstrated good judgment, so why shouldn't we rely on their judgment?

At the same time, Barksdale highlights the tremendous opportunity for all of us as individuals. Relying on data—indeed, expecting every conversation to be rooted in data—upends the traditional role of managers. It transforms them from being providers of intuition to facilitators in a search for truth, with the most useful facts being brought to bear on each decision. In a sense, every meeting becomes a Hegelian dialectic, with presenters providing a thesis and the folks in the room providing an antithesis, spurning opinion, questioning facts, and testing which decision is correct. The result is synthesis, a closer approximation of truth than if we had relied on mere pronouncements. One of the core principles of Google has always been "Don't politick. Use data."

As Hal Varian told me, "Relying on data helps out everyone. Senior executives shouldn't be wasting time debating whether the best background color for an ad is yellow or blue. Just run an experiment. This leaves management free to worry about the stuff that is hard to quantify, which is usually a much better use of their time."

We use data—evidence—to guard against rumor, bias, and plain old wrongheadedness. One way is by myth busting (with apologies to Adam Savage and Jamie Hyneman's superb television program *MythBusters*, where they test whether pop culture myths are true: "Is it possible to escape from Alcatraz Island?"[xxxvii] "If it's

---

[xxxvii] Yes.

raining, will you get wetter by running or walking?"[xxxviii]). Inspired by the show, we try to test myths within the company and debunk them wherever we can.

People make all kinds of assumptions—guesses, really—about how things work in organizations. Most of these guesses are rooted in sample bias. A textbook illustration of sample bias is Abraham Wald's work in World War II. Wald, a Hungarian mathematician, was a member of the Statistical Research Group (a group based at Columbia University that took on statistical assignments from the US government during the war). He was asked what the military could do to improve the survival rates for bombers. Wald reviewed the location of bullet holes on planes returning from bombing runs to determine where adding more armor would help. According to the National WWII Museum,[100] he created the diagram below. The darkened areas on the plane on the right show where the most holes were.

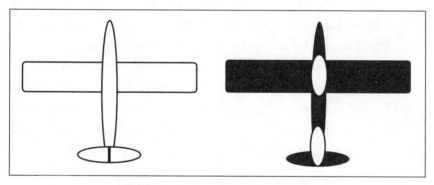

Wald's illustration of bomber damage.

Wald concluded, counterintuitively, that the cockpit and the tail needed the most reinforcement.

The sample he was looking at included only bombers that survived and had been shot up throughout the wings, nose, and

---

[xxxviii] Walking gets you wetter.

fuselage. Wald realized he was looking at a biased sample: The bombers that were shot in the cockpit and tail never made it back. That's where the bulletproofing was most needed.[xxxix]

Sample bias afflicts us all. For example, in 2010 our annual employee survey revealed that many engineering Googlers felt that Google didn't take firm enough action on poor performers. What was really happening was that on one team of ten people, nine were all looking at the same struggling performer and concluding that no one was doing anything to help them improve or fire them. They didn't see the other five teams of like size where no one was struggling, nor was it often known that managers and People Operations folks were working with the individual behind the scenes. This is classic sample bias, where someone is drawing conclusions based on the small, flawed sample that they happen to see. In this case, a respect for the privacy of the struggling person kept us from broadcasting what was happening, but I made a point to share with Googlers that we are constantly working behind the scenes. We also incorporated the actual data from all Googlers into training materials and talks related to performance management.

As a result, engineering Googlers have become significantly more positive about this issue, scoring 23 points (on a scale of 100) more favorably than before on the question of "In my work group, we deal effectively with low performers." Even better, Googlers are now explaining the dynamic to one another. In a recent email

---

[xxxix] This is also an example of survivorship bias, where you skew your analysis by considering only the survivors rather than the entire population. Analysts looking at the performance of start-up companies and hedge funds often make this error because they include only companies that are still around, and ignore those that fail or shut down along the way. This makes start-up and hedge-fund performance look rosier than it is. And, of course, relying too much on this book may also be an example of survivorship bias. There are certainly lessons to be learned from the illustrations in this book, but it's important to consider the lessons of failed companies as well. In People Operations we try to avoid survivorship bias in our analyses where possible. For example, we've tested some of our hiring practices by—in a double-blind fashion—hiring rejected candidates to see how they perform.

thread, one Googler was frustrated because of what he perceived to be a slacker performer whose shortcomings weren't being addressed. Another Googler chimed in to explain that the slacker probably was getting attention, but that the "lawyercats" [101] likely wouldn't allow People Operations to share the details with everyone. True!

Promotions are another area where myths arise. We announce promotions at Google with emails listing people's names and short biographies, but we're large enough that it's impossible to know everyone. So Googlers scan the list of names, naturally looking for the people they know so they can congratulate them. But they also make unconscious assumptions as they go. "I noticed Sally got promoted but Dave didn't. Must be because Sally works with our CFO." "Wow, look at all the people from the Android [mobile device] team that were promoted.... I hardly see anyone from Infrastructure [data centers]. I guess we only care about the user-facing stuff." Typical inferences are that you need to have much more senior people on your project to get promoted, since their voices presumably carry more weight; it helps to be in a "sexier" product area to get promoted; a single negative review can torpedo your odds of promotion; or projects based at our headquarters are more visible, and it's therefore easier to get promoted while working on those. The list goes on. Each year, when we surveyed Googlers they told us that the promotion process wasn't fair because of all the favoritism shown to certain offices, projects, and jobs.

All of these would be legitimate concerns, if they were correct inferences. But they're not.

When a Googler takes the time to check their observations with People Operations, we walk them through the data. But most Googlers don't ask—and we wouldn't have time to answer every question anyhow. And not only do we constantly have new hires come in and draw the same wrong conclusions, but even Googlers who've been around awhile sometimes still preserve a (perhaps healthy) level of skepticism about our assurances. After all, doesn't the HR team work for management? Don't they have an interest in quieting the masses?

Enter Brian Ong and Janet Cho. Brian leads the team that tracks and measures every part of our hiring process, but several years ago he was a member of our People Analytics team, charged with making sure we supplemented our judgment with facts. Janet is our VP in charge of all people issues related to our major product areas, such as search, advertising, data centers, and Gmail. They decided that a more effective and long-lasting approach would be simply to share all the promotion data with Googlers. They crunched the numbers, organized a series of talks, recorded them so people could watch them later, and built a site to share all the data. It turns out:

- Working with much more senior people has only a small effect. Fifty-one percent of all people nominated for promotion were promoted. For those who worked with much more senior people, the promotion rate was 54 percent. A little higher, but not much.

- The product area doesn't matter. There are occasional differences of a few percentage points in one year or another, but in general your odds of promotion are the same no matter what you work on.

- Bad feedback doesn't hurt you. In fact, almost every person who gets promoted has constructive feedback in their promotion materials. What ruins someone's chances is evidence of something seriously wrong, such as poorly organized or consistently buggy code. The other warning sign is a conspicuous absence of information. A promotion packet that has no constructive feedback is actually a warning sign to review committees. Promotion candidates shouldn't be afraid to solicit and receive less-than-glowing feedback, since it won't derail them, and it will give them explicit coaching on how to improve. It turns out that when you present people with reality, they want to get better.

- Where your project is based doesn't affect your ability to get promoted. For example, the promotion rate at our headquarters in Mountain View is virtually the same as elsewhere.

The site periodically gets updated with the latest facts and any new analyses that have been requested. It's a lot of work, but essential to demonstrate that our processes are unbiased. It would have been easy to keep asserting that the process worked. But far better to bust the myths once and for all with facts, and then make those facts freely available to anyone.

We test ourselves and Google products frequently to make sure our decisions are fact based. We want to bolster good ideas while weeding out bad ones, which in turn gives us more room to experiment freely with the most promising ones. In 2010, for example, we made 516 improvements to how Google Search works. One example of a major improvement was code-named Caffeine, which made our results 50 percent "fresher" than before. Google doesn't search the entire Web every time someone types a search query. Instead, we search in advance, with different sites and pages prioritized according to relevance, quality, and so on, and then we index those sites so that a query gets an answer virtually instantly. Caffeine meant that we were indexing 50 percent faster. At launch, it processed hundreds of thousands of Web pages in parallel each second. If printed, that would be a stack of pages three miles high every second.[102]

Before implementing any of the improvements, we tested them to make sure they work. We used A/B testing, showing evaluators two sets of search results side by side, then watching their behavior and getting their feedback on which results were better. An easy illustration would be testing whether ads with blue backgrounds or red backgrounds get more clicks from users. It seems like a small question, but it's a big one if you're Coca-Cola or Pepsi. We also use one-percent tests, where we roll out a change to one percent of users to see what happens before implementing the change for billions of users. In 2010 alone, we conducted 8,157 A/B tests and more than 2,800 one-percent tests. Put another way, every single day in 2010 we ran more than thirty experiments to uncover what would best serve our users. And this was just for our search product.

We take the same approach on people issues. When we implemented our Upward Feedback Survey (a periodic survey about manager quality—more on this in chapter 8), we ran an A/B test to see if Googlers were more likely to give feedback to their managers if the email announcing the survey was signed by an executive or a generic "UFS Team" email alias. We saw no difference in response rates, so we opted to use the generic alias, simply because it's easier to write one email than to ask each executive to write one of his or her own.

Almost any major program we roll out is first tested with a subgroup. I remember when we crossed twenty thousand employees and I was first asked if it bothered me that Google was now indisputably a big company. "We always worry about culture," I responded, "but the virtue of big is that we can run hundreds of experiments to see what really makes Googlers happier." Every office, every team, every project is an opportunity to run an experiment and learn from it. This is one of the biggest missed opportunities that large organizations have, and it holds just as true for companies made up of hundreds, not thousands. Too often, management makes a decision that applies unilaterally to the entire organization. What if management is wrong? What if someone has a better idea? What if the decision works in one country but not another? It's crazy to me that companies don't experiment more in this way!

Why not carve out ten or fifty or a hundred people and try something different? Or try something first with a small group? As they used to say, "If you're not careful, you may learn something before you're done."

Find ways for people to shape their work and the company

In addition to stripping leaders of the traditional tools of power and relying on facts to make decisions, we give Googlers uncommon freedom in shaping their own work and the company. Google isn't the first to do so. For over sixty-five years, 3M has offered its employees 15 percent of their time to explore: "A core belief of 3M is that creativity needs freedom. That's why, since about 1948, we've encouraged our employees to spend 15% of their working time on their own projects. To take our resources, to build up a unique team, and to follow their own insights in pursuit of problem-solving." [103] Post-it Notes famously came out of this program, as did a clever abrasive material, Trizact, which somehow sharpens itself as it's used.

Our version is 20 percent time, meaning that engineers have 20 percent of their week to focus on projects that interest them, outside of their day jobs but presumably still related to Google's work (at Google that still covers a lot of territory). Outside of engineering, we don't formally label projects as 20 percent time, but Googlers often find time for their own side projects, whether it was salesman Chris Genteel deciding to help minority-owned businesses get online (which eventually turned into a full-time job for him) or Anna Botelho, a former competitive ballroom dancer and member of our real estate team, enlisting other Googlers to teach dance classes at Google.

Caesar Sengupta, a VP of Product Management on our Chrome team, had a day job in 2009 running Google Toolbar and Desktop, downloadable versions of our products that sat inside your browser or on your desktop computer. When the Chrome team started building their browser, Caesar and a few engineers wondered what would happen if you applied Chrome's design to operating systems, the programs that tell your phone, tablet, or computer how to work. At the time, it could take five minutes or longer for a computer to boot up, in part because it was still checking for antiquated hardware that no one used anymore, like floppy disk drives. Caesar and the team started an informal 20-percent-time project to do better. They stripped out all the unnecessary steps, built on the Chrome browser platform, and created their first prototype Chromebook laptop. It booted in eight seconds.

Utilization varies in practice, with some individuals focusing virtually 100 percent on side projects and many others not having any side projects at all. Some joke that it's really "120 percent time," where work is done after the day job rather than instead of it. More typically, a successful project starts with 5 or 10 percent of someone's time, and as it demonstrates impact it consumes more and more time (and attracts more and more volunteers) until it becomes a formal product.

The use of 20 percent time has waxed and waned over the years, humming along at about 10 percent utilization when we last measured it. In some ways, the idea of 20 percent time is more important than the reality of it. It operates somewhat outside the lines of formal management oversight, and always will, because the most talented and creative people can't be forced to work.

Ryan Tate of *Wired* wrote the best summary of it I've seen:[104]

Here is what [20 percent time] is not: A fully fleshed corporate program with its own written policy, detailed guidelines, and manager. No one gets a "20 percent time" packet at orientation, or pushed into distracting themselves with a side project. Twenty percent time has always operated on a somewhat ad hoc basis, providing an outlet for the company's brightest, most restless, and most persistent employees—for people determined to see an idea through to completion, come hell or high water.

For example, engineer Paul Buchheit worked on Gmail for two and half years before he finally persuaded company brass, who worried about stretching Google too far beyond search, to launch the thing.

Googlers don't restrict themselves to creating products. They also involve themselves in deciding how we run the company. A few years back, we gave a group of thirty engineers anonymous performance and pay data for everyone in engineering, and allowed them to shape how bonuses would be allocated. They wanted the system to be more meritocratic. For example, imagine two engineers performing at the same level, but one was better able to negotiate his salary when he joined Google. Let's say he had a salary of $100,000 per year, and another engineer, who didn't think to negotiate, had a salary of $90,000. Since they performed at the same level, both received a 20 percent bonus. That's not fair, argued the engineers, because the first person received $20,000 in bonus, while the second had contributed the same impact but was paid only $18,000. So, at their request, we changed the basis for bonus calculation from actual salaries to the median salary of all people in that job. That ensured that both people received a bonus commensurate with their impact.

As an aside, this is a very real issue in most companies. There is a well-documented average salary difference between men and women. One source of this is a difference in men's and women's propensity to negotiate when they are being hired. For example, Linda Babcock of Carnegie Mellon University and author Sara Laschever reported that starting salaries for male MBA graduates from Carnegie Mellon were higher than for females, largely because men were more likely to ask for higher salaries. Fifty-seven percent of men negotiated, compared to 7 percent of women.[105] In part thanks to input from Googlers, our pay systems are built with a goal of eliminating this kind of structural bias and inequity.

But we didn't always approach compensation, utilization, hiring, or other people issues so analytically. Back in 2004, when we had about 2,500 Googlers, Larry and Sergey felt we were becoming so big that they couldn't get an intuitive sense of how happy people were just by walking around and talking to people they knew. Their solution: Stacy Sullivan should interview everyone and find out.

Stacy countered by proposing the Happiness Survey, but far less than half of the company participated. And the engineers, believing that they could design a better one, launched their own competing survey. Its name? The Ecstasy survey, since they of course had to set the bar even higher. The Ecstasy survey addressed the specific needs of engineers—for example, it was the only survey to ask about 20 percent time usage—and was initially more credible with other technical staff (because, after all, an engineer designed it!).

Until 2007 the surveys ran in parallel, but their utility was limited because, with different sets of questions, we couldn't ever make comparisons across the entire company. Michelle Donovan, who would later work on Project Oxygen, wanted to find a better way. She spent the next year partnering with engineers, salespeople, and everyone else to develop a survey that would capture the interests of all Googlers and would also be scientifically robust and measurable over time. Googlegeist was born.

Googlegeist, which means "the spirit of Google" and was—no surprise—chosen by employees, is an annual survey of our fifty thousand–plus Googlers. It is our most powerful single mechanism for enabling our employees to shape the company. Googlegeist asks about a hundred questions each year, scored on a five-point scale from "strongly disagree" to "strongly agree," supplemented with several free-response questions.

Gooooglegeist
Our Annual Survey

Section 1: Me

The Google-wide portion of the survey is organized into four sections. This first section asks you about topics that relate to your individual experience as a Googler.

If you prefer not to answer a question, don't know the answer, or feel that the question doesn't apply to you, please select "N/A."

	Strongly disagree	Disagree	Neutral	Agree	Strongly agree	N/A
	○	○	○	○	○	○
	○	○	○	○	○	○
	○	○	○	○	○	○
	○	○	○	○	○	○
	○	○	○	○	○	○
	○	○	○	○	○	○
	○	○	○	○	○	○

0% 25% 50% 75% 100%

[Previous] [Next]

Click "Next" to save your answers.

The first page of the 2014 Googlegeist survey. © Google, Inc.

We change 30 to 50 percent of the questions each year, based on what issues are most pressing, but retain the rest so we can track changes in the company over time. Roughly 90 percent of Googlers participate each year.

Because we want Googlers to be honest, there are two ways to submit your responses: confidentially or anonymously. "Confidentially" means that your name is stripped out, but other data that help us analyze the company are left in—for example, your location, job level, and product area, so we would know that a Googler is a female manager at YouTube in San Bruno, California, but not who she is specifically. The only team that ever sees that data—without names of course—is the core Googlegeist team, and we never report results in a way that allows any individual to be identified. An anonymous submission goes a step further, including no personally identifying information unless the respondent chooses to add it.

Googlegeist is distinctive because it is written not by consultants but by Googlers with PhD-level expertise in everything from survey design to organizational psychology, all results (both good and bad) are shared back with the entire company within one month, and it is the basis for the next year of employee-led work on improving the culture and effectiveness of Google. Every manager with more than three respondents gets a report, dubbed MyGeist. This report—which is actually an interactive online tool—allows managers to view and share personalized reports that contain a summary of the Googlegeist scores for their organizations. Whether their team is three or thirty Googlers strong, managers get a clear sense of how they're doing according to their teams. With one click, managers can choose to share with just their direct teams, with their broader organizations, with a customized list of Googlers, or even with all of Google. And most do.

Example of personalized Mygeist report from the 2014 Googlegeist survey.
Data is illustrative. © Google, Inc.

There's a virtuous cycle here: We take action on what we learn, which encourages future participation, which then gives us an ever more precise idea of where to improve. We enable this cycle by defaulting to open: The reports of any vice president with a hundred or more respondents are automatically published to the entire company, a thought that must terrify the CEO we met in chapter 2 who was too timid to have an unscripted Q&A with his employees. At the same time, employee responses are anonymous (to eliminate sycophancy) and managers' results are not factored into performance ratings or pay decisions. We want employees to be scathingly honest, and managers to be open to improvement rather than defensive.

Critically, Googlegeist focuses on outcome measures that matter. Most employee surveys focus on engagement,[106] which as Prasad Setty explains, "is a nebulous concept that HR people like but doesn't really tell you much. If your employees are 80 percent engaged, what does that even mean?"[xl] The Corporate Executive Board found that "The meaning of employee engagement is ambiguous among both academic researchers and among practitioners.... [The] term is used at different times to refer to psychological states, traits, and behaviors as well as their antecedents and outcomes."[107] Engagement doesn't tell you precisely where to invest your finite people dollars and time. Do you increase it by focusing on health programs? On manager quality? On job content? There's no way to know.

Googlegeist instead focuses on the most important outcome variables we have: innovation (maintaining an environment that values and encourages both relentlessly improving existing products

---

[xl] As a former consultant, I can tell you that many tout engagement as a panacea. They measure engagement through a short questionnaire, typically including statements like: "I have a best friend at work," "In the last seven days, I have received recognition or praise for doing good work," or "My supervisor, or someone at work, seems to care about me as a person."

My chief HR officer friends tell me that engagement surveys fail to tell them how to improve. If your scores are low, do you raise them by somehow convincing more employees to be best friends? Or, if profits are low, is the best fix to start praising people more? We do measure some similar topics at Google (along with dozens more), but don't merge them into a single all-encompassing construct like engagement. We see better results by instead understanding very specific areas like career development or manager quality.

and taking enormous, visionary bets), execution (launching high-quality products quickly), and retention (keeping the people we want to keep). For example, we have five questions that predict whether employees are likely to quit. If a team's responses to those five questions fall below 70 percent favorable, we know more people will leave during the following year unless we intervene. If scores fall below 70 percent for just one question, the issue is identified and Googlers and leaders partner with their People Operations colleagues to improve that team's experience (though note that results are never tied back to any individual Googler). We measure many other outcomes, such as our pace of execution and our culture, but above all we want to continue launching cool new stuff, and we want to ensure that the people we've worked so hard to hire remain at Google for a long time.

The effect has been profound. We've been able to anticipate areas where attrition might go up and keep employee turnover consistent and low, through good times and difficult ones. Googlers continue to feel the company is innovative and that they can contribute to our mission. But compared to five years ago, Googlers are 20 percent more confident that their career goals can be met at Google, 25 percent happier with the pace of decision-making (which can slow dramatically as bureaucracy creeps into larger organizations), and feel 5 percent more treated with respect (it's hard to improve much when some baseline scores are already above 90 percent).

At the same time, the survey reveals we have work to do in some areas of well-being, specifically in Googlers' ability to detach from work during nonwork times. And so we're trying to get better. Our Dublin office responded by creating a program called Dublin Goes Dark, where everyone was encouraged to leave work at 6:00 p.m. and stay offline. They even had drop-off locations to turn in laptops, to ensure people weren't sneaking a peek at email before going to bed. The experiment worked. What started as a People Operations–only effort in 2011 became a Dublin-wide event by 2013, with our entire Dublin office of more than two thousand people participating. Helen Tynan, our People Operations leader in Ireland, reported: "I had a stack of laptops in my office, and quite a buzz (lots of giggling) as people dropped them in....Lots of people chatted the next day about what they had done, and how long the evening had seemed with lots of time for doing things."

Googlegeist has also been a source of major changes we've made to the company, among them Project Oxygen—which transformed how we think about management at the company and which I'll detail in chapter 8—and a sweeping change to our pay philosophy that we implemented in 2010, which included a 10 percent salary increase for everyone in the company. Until then, our salaries had been quite low, but as Googlers started buying homes and raising families, having a higher fixed wage became more important. We saw the level of satisfaction with salaries shrink over time, and acted. (Though, sadly, both Larry and Sergey, who take only $1 per year in salary, declined my offer to raise their salaries by 10 percent...to $1.10 per year.)

But in a company where the glory has historically been in launching new products, Googlegeist revealed that we were overlooking some fundamental needs.

In 2007 and 2008, tech Googlers didn't feel there was enough recognition given to people doing less flashy but important work. For example, Google's engineers contribute to our codebase simultaneously. Let that sink in for a moment. Thousands of people are changing how Google's products work all the time, and all at the same time. Small duplications and inefficiencies can add up quickly and make products slow, overly complex, and buggy.

"Code health" refers to maintaining the overall sustainability and scalability of code to minimize this problem. ("Scalability" is tech industry jargon—it's what we mean when we talk about being able to take a small solution and make it work for the entire world.

"Scale" means something works equally well for a hundred users and for one billion users.) It requires regularly researching and creating techniques to reduce complexity, and integrating an ethos of simplification into our code development process. Googlegeist told us that we weren't paying enough attention to this, or rewarding it sufficiently. Rewards were accruing to those who generated the most volume of code, rather than the highest-quality code.

We could have set company-wide goals for code health, or created new jobs entirely focused on checking others' work. Or our CEO could have just mandated that everyone had to focus on code health for the next month.

Instead, a group of engineers self-organized and decided to tackle the problem after it was highlighted in Googlegeist. First, they worked to improve recognition of the importance of code health through education and publicity, including Tech Talks, articles on our intranet, and emails sent to our popular "eng-announce" internal listserv explaining why code health matters. They asked revered technical executives like Alan Eustace to work mentions of code health into public talks and emails about performance management and promotions. Second, the engineers partnered with People Operations to ensure code health became an essential component of performance calibration sessions and promotion committees, and incorporated questions into Googlegeist to check on progress annually. Third, they worked to build tools to automatically review code health. For example, our Munich code health group developed a tool that automatically detects dead code in C++ and Java. Spotting and excising dead code makes your programs run faster and more reliably. Finally, the engineers created the Citizenship Awards, an opportunity for Googlers who contribute to healthy code to be recognized by their peers and leaders, all of whom benefit from their work.

Four years later, engineers are 34 percent more confident that time spent improving code health will be rewarded. More important, they have started reporting small but measurable improvements in their own productivity, both because their project team's codebase was getting stronger and because the systems on which they depended outside of their team were also improving.

Googlegeist regularly reveals opportunities to improve elsewhere at Google as well. In our sales organization, we saw a dip in career development satisfaction among new graduate Googlers who work with small businesses (such as a downtown boutique or a tapas joint in Brooklyn). The Googlers themselves, partnering with People Operations, designed a pilot program in Europe focused on rotating Googlers through various jobs, role-specific business and product training, a two-year personal development plan, and building networks across the world.

The first cohort to go through the program saw an 18-point improvement (out of a possible 100) in career development satisfaction, and an 11-point improvement in retention indicators. Based on the successful pilot, the program has rolled out to almost eight hundred Googlers across the globe.

We also run "quick hit" programs periodically, focused on more targeted issues. Bureaucracy Busters fixes all the annoying little impediments that make life exasperating. For example, we no longer use paper receipts for expense reporting: Take a picture and send it in. The Waste Fix-It is about stamping out small practices that waste money, like having more printers than we need. We ask Googlers to suggest fixes that would benefit lots of their colleagues and that we can implement within two or three months. In 2012, we received 1,310 ideas and over 90,000 votes on them. The top twenty were implemented. Many weren't revelatory: Stop sending paper paychecks; make sure mandatory ethics and compliance training doesn't happen at year-end, right as everyone is also doing new product roadmaps and budgeting for the next year; develop structured interviewing tools to automatically give you reliable interview questions, so you're not forced to make up your own wacky ones; get more corporate housing in cities like New York or London where hotels are expensive. But they were all fixes for tiny, irritating practices that could add up to the company feeling stodgy and slow.

Expect little from people, get little. Expect a lot...

Some will argue that this kind of mass empowerment leads to anarchy, or to a situation where, since everyone's opinion is valued, anyone can object and derail an effort—an environment where ten thousand people can say no but no one can say yes. The reality is that every issue needs a decision maker. Managed properly, the result of these approaches is not some transcendent moment of unanimity. Rather, it is a robust, data-driven discussion that brings the best ideas to light, so that when a decision is made, it leaves the dissenters with enough context to understand and respect the rationale for the decision, even if they disagree with the outcome.

This approach almost always works. When it fails, there's a simple rule to follow: Escalate to the next layer of the company and present the facts. If they can't decide, escalate again. In our company, eventually Larry Page will break the tie.

This advice may seem out of place, given that I've just devoted much of this chapter to explaining why managers should not have authority. But hierarchy in decision-making is important. It's the only way to break ties and is ultimately one of the primary responsibilities of management. The mistake leaders make is that they manage too much. As Olivier Serrat of the Asian Development Bank wrote, "Micromanagement is mismanagement.... [P]eople micromanage to assuage their anxieties about organizational performance: they feel better if they are continuously directing and controlling the actions of others—at heart, this reveals emotional insecurity on their part. It gives micromanagers the illusion of control (or usefulness). Another motive is lack of trust in the abilities of staff—micromanagers do not believe that their colleagues will successfully complete a task or discharge a responsibility even when they say they will."[108]

Instead, decisions should be made at the lowest possible level of an organization. The only questions that should rise up the org chart are ones where, Serrat continues, "given the same data and information," more senior leaders would make a different decision than the rank and file.

You don't need to have Google's size or analytical horsepower to unleash the creativity of your people. As a leader, giving up status symbols is the most powerful message you can send that you care about what your teams have to say. When I worked for a fifty-person company, the greatest thing the chief operating officer, Toby Smith, did to make me feel like an owner was share an office with me. By watching him each day I learned about our business and how to connect with people (he always answered the phone with a startlingly warm "How ARE you?"), as well as benefited from his office tips (when you buy dress shoes, buy two identical pairs and rotate them so they don't wear out from daily use[xli]). There are free survey tools, including one built into Google Sheets, that let you ask employees how they are feeling and what they'd like to do differently. Using small pilots allows the most vocal employees to grapple with the complexities of a situation. It's easy to complain from the sidelines. Being charged with implementing your ideas is often much harder and can moderate more extreme and unrealistic perspectives.

All this adds up to happier people generating better ideas. The truth is that people usually live up to your expectations, whether those expectations are high or low. Edwin Locke and Gary Latham, in their 1990 book *A Theory of Goal Setting & Task Performance*, showed that difficult, specific goals ("Try to get more than 90 percent correct") were not only more motivating than vague exhortations or low expectations ("Try your best"), but that they actually resulted in superior performance. It therefore makes sense to expect a lot of people.

When I was at McKinsey & Company, I had a manager named

[xli] I still have the two pairs of oxfords I bought in 1994.

Andrew who expected perfection in the market analyses I prepared for clients. But he didn't micromanage me by telling me how to write each page or how to do my analyses. Andrew set our expectations higher.

In 1999 we were serving a financial services company and doing one of the first e-commerce projects our firm had ever done. (Remember "e-commerce"?) I brought a draft report to him and instead of editing it, he asked, "Do I need to review this?" I knew deep down that while my report was good, he would surely find some room for improvement. Realizing this, I told him it wasn't ready and went back to refine it further. I came back to him a second time, and a second time he asked, "Do I need to review this?" I went away again. On my fourth try, he asked the same question and I told him, "No. You don't need to review it. It's ready for the client."

He answered, "Terrific. Nice work." And sent it to the client without even glancing at it.

If you expect little, that's what you'll get. Richard Bach, the author of the 1970s bestselling novel *Jonathan Livingston Seagull*, later wrote in *Illusions*, "Argue for your limitations, and sure enough, they're yours." [109] Managers find many reasons not to trust their people. Most organizations are designed to resist change and enfeeble employees. When I tell CEOs that many Googlers can nominate themselves for promotion, or that they can ask our CEO any question, the most common response is that it sounds great in theory but would never work at their company: People won't focus on their real work; gathering all those facts would just slow us down (!); the lawyers won't let us do it; people (the very ones that are their "most important asset") won't make good decisions; I like my special parking space....

But it does work. You just need to fight the petty seductions of management and the command-and-control impulses that accompany seniority. Organizations put tremendous effort into finding great people but then restrict their ability to have impact on any area but their own tasks.

As a manager, it can be terrifying to let go of the reins. After all, your career is on the line if something goes wrong. And you were put in this job because you're supposedly the best suited to lead.

It's been hard for me to let go as well. But I noticed a funny thing. Each week Larry asks that People Operations, like other areas of Google, write a short brief about the week before for the entire management team to review. It's not an evaluation, just a way for everyone to know what's going on. At first, I wrote it myself each week. Then I asked folks on my team to write it, and I'd review and edit. Eventually, I asked Prasad to review it and send it without my input.

On the one hand, no big deal: It's just a memo. On the other hand, this is the only regular representation to our CEO of what the entire People Operations team is doing. In letting go, I gave up a little bit of control. But I got back valuable time to focus on other pressing issues. And Prasad got the opportunity to take on something new.

What managers miss is that every time they give up a little control, it creates a wonderful opportunity for their team to step up, while giving the manager herself more time for new challenges.

Pick an area where your people are frustrated, and let them fix it. If there are constraints, limited time or money, tell them what they are. Be transparent with your people and give them a voice in shaping your team or company. You'll be stunned by what they accomplish.

WORK RULES... FOR MASS EMPOWERMENT

- ☐ Eliminate status symbols.
- ☐ Make decisions based on data, not based on managers' opinions.
- ☐ Find ways for people to shape their work and the company.
- ☐ Expect a lot.

Why Everyone Hates Performance Management, and What We Decided to Do About It

Improve performance by focusing on personal
growth instead of ratings and rewards

I n the *Simpsons* episode called "The PTA Disbands," the
teachers of Springfield Elementary School go on strike, pro-
testing the lack of spending on salaries, school supplies, and
food. With the school shut down, the children are left to their own
devices. Some respond by playing video games all day, others by
playing pranks. Second grader Lisa Simpson panics:

> *Lisa:* But without state-approved syllabi and standardized test-
> ing, my education can only go so far.
> *Marge Simpson (her mother):* Honey, maybe you should relax a little.
> *Lisa:* Relax? I can't relax! Nor can I yield, relent or...only two
> synonyms? Oh my God! I'm losing my perspicacity!
> *Marge:* Well, it's always in the last place you look.

After a few days, it gets worse for Lisa:

> *Lisa:* Look at me! Evaluate and rank me! I'm good, good, good,
> and oh so smart! Grade me!

Marge: I'm worried about the kids, Homie. Lisa's becoming very obsessive. This morning I caught her trying to dissect her own raincoat.

There's a little bit of Lisa Simpson in all of us. As children we line up from shortest to tallest. We're graded and told we are outstanding, satisfactory, or need improvement. As we get older we are ranked in our classes and take tests that compare us to national averages. We apply to university, mindful of how each school is ranked. Our first twenty years are spent being compared to others.

It's no wonder, then, that as adults we re-create those same conditions when we design our work environments. It's what we know.

And Google has been no different. We need people to know how they're doing, and we've evolved what might at first seem like a zanily complex system that shows them where they stand. Along the way, we learned some startling stuff. We're still working on it, as you'll see, but I feel pretty confident we're headed in the right direction. And with any luck I can save you some of the headaches and missteps we had along the way.

Throwing in the towel

The major problem with performance management systems today is that they have become substitutes for the vital act of actually managing people. Elaine Pulakos, a PhD psychologist from Michigan State University and now president of PDRI, a top consulting firm in this area, observed that "[a] significant part of the problem is that performance management has been reduced to prescribed, often discrete steps within formal administrative systems....Although formal performance management systems are intended to drive... the day-to-day activities of communicating ongoing expectations, setting short-term objectives, and giving continual guidance... these behaviors seem to have become largely disconnected from the formal systems."[110]

In other words: Performance management as practiced by most organizations has become a rule-based, bureaucratic process, existing as an end in itself rather than actually shaping performance. Employees hate it. Managers hate it. Even HR departments hate it.

The focus on process rather than purpose creates an insidious opportunity for sly employees to manipulate the system. Don, a sales leader I once worked with (though that's not his real name), would start visiting my office in the three months before we would do executive ratings and bonuses. Every October, he would start laying a foundation. "It's looking like a tough year but the team is working hard to get through it," Don would report. By November there was an update: "The sales guys are doing better than it looks like, fighting against this tough economy." By December we'd be into the details: "The small-business team is coming in at 90 percent, but boy did that team fight like heroes to bring in those deals. And by the way, I can't believe the crazy goals that were set back in January. Those were impossible!"

I didn't realize the game Don was playing until one year when we decided to pay bonuses one quarter later than usual, but didn't tell Don. He showed up six months early to pre-negotiate. Frankly, the fact that he was always working the angles made him a great salesman, but the episode also taught me the degree of gaming that went into the system.

In fact, no one is happy about the current state of performance management. WorldatWork and Sibson Consulting surveyed 750 senior HR professionals and found that 58 percent of them graded their own performance management systems as C or worse. Only 47 percent felt the system helped the organization "achieve its strategic objectives," and merely 30 percent felt that employees trust the system.[111]

The response in vogue today is to surrender.

Adobe, Expedia, Juniper Networks (a computer hardware manufacturer), Kelly Services (a temporary worker agency), and Microsoft have all eliminated performance ratings. Adobe's abandonment of ratings is telling:

> During a trip to India, [Adobe's] sleep-deprived [chief HR officer Donna] Morris was being interviewed for an article in *The Economist*. Feeling "edgier than normal," Morris spoke openly about her growing desire to abolish the performance review. Scrambling to get in front of the story, Morris worked with Adobe's communications department to quickly write a blog entry on the subject that was posted on the company's intranet. Employees devoured the post, making it one of the most-read pieces in the history of Adobe's intranet. Across the company, they engaged in a lively discussion about their dissatisfaction with the review process. According to Morris, the underlying message that emerged was that employees were "disenchanted about what they believed to be a lack of recognition for their contributions." For Morris, the necessary course of action became obvious.
>
> "We came to a fairly quick decision that we would abolish the performance review, which meant we would no longer have a one-time-of-the-year formal written review," says Morris. "What's more, we would abolish performance rankings and levels in order to move away from people feeling like they were labeled."
>
> In place of the traditional performance review, Adobe introduced The Check-In—an informal system of ongoing, real-time feedback—in the summer of 2012.[112]

Intuitively, this sounds appealing. Employees are unhappy, so throw out the system they don't like. Simple.

And isn't receiving feedback in real time better than waiting for a year?

But there's no evidence that the systems people use to do this work either. The academic research suffers from inconsistent measurement, where "real time" can mean anything from "immediately" to "days later." Most real-time feedback systems quickly turn into "attaboy" systems, as people only like telling each other nice things. And how often are your comments structured in a way that actually causes behavior to change? Saying "Great job in that meeting" is far more common than "I saw that you noticed when the customer pushed back from the table and seemed to lose interest, and you responded by asking if they had any concerns. You did a great job of re-engaging them. You should continue to pay close attention to body language in meetings."

It's far, far easier and more common to stick with vague pleasantries.

Setting goals

Even at Google, our system was far from perfect. Satisfaction with performance management was consistently one of the lowest-rated areas in our annual Googlegeist study. In early 2013, only 55 percent of Googlers viewed the process favorably. Better than the 30 percent favorability seen in other companies, but still pretty awful. The two primary complaints were that it took too much time and the process wasn't transparent enough, which raised concerns about fairness. So what were we doing right that made our employees twice as happy with the system as employees elsewhere—but still not happy enough? And what were we doing wrong?

Google's performance management system has always started with goal setting. In the early 2000s, Google board member John Doerr introduced us to a practice he had seen Intel use with much success: OKRs, or Objectives and Key Results. The results must be specific, measurable, and verifiable; if you achieve all your results, you've attained your objective. For example, if the objective is to improve search quality by x percent, key results that contribute to that would be better search relevance (how useful the results are to the user), and latency (how quickly the results show up). It's important to have both a quality and an efficiency measure, because otherwise engineers could just solve for one at the expense of the other. It's not enough to give you a perfect result if it takes three minutes. We have to be both relevant and fast.

We deliberately set ambitious goals that we know we won't be able to achieve in all cases. If you're achieving all your goals, you're not setting them aggressively enough. Astro Teller,[xlii] who oversees Google[x], our team that developed Glass (an eyeglass-mounted computer with a viewscreen the size of your fingernail) and our self-driving cars, describes it this way: "If you want your car to get fifty miles per gallon, fine. You can retool your car a little bit. But if I tell you it has to run on a gallon of gas for five hundred miles, you have to start over." We don't set all our goals quite that aggressively, but there's wisdom in his approach. As Larry often points out, "If you set a crazy, ambitious goal and miss it, you'll still achieve something remarkable."

So at the beginning of each quarter, Larry sets OKRs for the company, triggering everyone else to make sure their own personal OKRs roughly sync with Google's. We don't let the perfect be the enemy of the good. Once you see the company's goals, it's easy enough to compare them to your own. If they're wildly out of step, either there's a good reason or you refocus. In addition, everyone's OKRs are visible to everyone else in the company on our internal website, right next to their phone number and office location. It's important that there's a way to find out what other people and teams are doing, and motivating to see how you fit into the broader picture

---

[xlii] Astro isn't his given name. His parents named him Eric, but it got a flat-top haircut after spending a year in Berkeley, California, father, physicist Edward Teller. His friends thought his head looke(and the name stuck. Any resemblance between his name and occu coincidental.

of what Google is trying to achieve. Finally, Larry's OKRs, followed by his quarterly report on how the company has performed, set the standard for transparency in communication and an appropriately high bar for our goals.

On the topic of goals, the academic research agrees with your intuition: Having goals improves performance.[113] Spending hours cascading goals up and down the company, however, does not.[114] It takes way too much time and it's too hard to make sure all the goals line up. We have a market-based approach, where over time our goals all converge, because the top OKRs are known and everyone else's OKRs are visible. Teams that are grossly out of alignment stand out, and the few major initiatives that touch everyone are easy enough to manage directly. So far, so good!

Measuring performance

Until 2013, every Googler received a performance rating score at the end of each quarter. The 41-point rating scale ran from 1.0 (awful) to 5.0 (astounding). Roughly speaking, below 3.0 meant you occasionally or consistently missed expectations, 3.0 to 3.4 meant you met them, 3.5 to 3.9 meant you exceeded, 4.0 to 4.4 was "strongly exceeded," 4.4 to 4.9 indicated "approaching astounding performance," and a 5.0 was "astounding." The average rating was between 3.3 and 3.4, and if someone had an average of 3.7 or higher for a few quarters, they were often promoted. Nothing revolutionary here.

The science on rating systems is inconclusive.[115] There's no strong evidence to suggest that having three or five or ten or fifty rating points makes a difference. Our 41-point scale came out of our engineering DNA. It felt satisfyingly precise to be able to draw a distinction between a 3.3 and a 3.4 performer. And when you averaged ratings over many quarters, you could calculate ratings finely enough to sort a 3.325 from a 3.350 performer. If you consider that ratings went out to three decimal places, we in reality had a 4,001-point rating system! We developed deliciously complicated formulae to make sure that if your rating was a hair higher than someone else's, you'd be rewarded with a slightly higher raise. But it didn't matter. For all the time we spent on assigning ratings, when it came time to set raises or bonuses, managers or later reviewers changed the pay outcomes two-thirds of the time. Our managers were spending thousands of hours every three months assigning ratings that were ludicrously precise but that weren't an accurate basis for determining pay.

The same goes for measuring performance four times per year. We started this practice both because in the years of torrential growth it helped us manage what people were doing, and because we wanted to make sure our assessments of people matched reality as closely as possible at all times. Yet we found that we were spending up to twenty-four weeks each year either assigning ratings, calibrating ratings (I'll explain what that means in a few pages—it's important), or communicating ratings. Some managers liked the frequency, arguing that it forced them to check in on people whose performance had suddenly gotten worse. But it was a crutch. There was no rule stopping them from checking on those people absent the reviews, and it sure seemed like a waste to have to review fifty thousand people so that you could find the five hundred who were struggling.

We spent most of 2013 exploring whether there was a better way. We examined alternatives ranging from having no job levels to having eight hundred job levels, so that almost everyone could have the morale boost of being promoted almost every quarter. We looked at assessing performance annually, quarterly, monthly, and in real time. We considered ratings systems with three points and fifty points. We debated whether we should label each performance category with a number or a descriptor, even kicking around the idea of meaningless descriptors so that people wouldn't focus on the labels. I even suggested a few that included rating levels like Aquaman, red triangle, or mango:

STRONGLY EXCEEDS MANGO

Image by Paul Cowan.[116]

Googler Paul Cowan created this graphic to illustrate
one of my rating proposals.

The point of the system was to get people to ignore it by using meaningless names. Of course, people would eventually assign meaning to the labels. And they'd probably start off by assuming Aquaman was the worst. (He always seems to lose out to the cooler superheroes.) We assembled steering committees, advisory committees, and even put some questions to a popular vote.

Ultimately, three things were clear:

1. Consensus was impossible. In the absence of clear evidence, everyone became an expert and there were constituencies arguing for every possible variation. People had strong opinions about questions like whether five or six performance categories are best. Even when making changes to the least popular process at Google, it was impossible to find a solution that made everyone happy. It seemed that even though

many people disliked the current system, they disliked every other option even more!

2. People took performance management seriously. For example, we polled Googlers about what to label our performance categories, and more than 4,200 votes were cast. The clearest trend was a desire for seriousness and clarity, not whimsy.

3. Experimentation was vital. In the absence of external evidence, we had to develop our own, working with the leaders of each part of Google to help them test their ideas. At YouTube they tried sorting everyone into a rank order of most to least effective, regardless of level, and found that one of the two most effective people was a mid-level employee, who was then rewarded with one of the biggest stock grants at YouTube. And while that specific person's award wasn't made public, everyone knew such reversals of level and pay were happening.[xliii] Elsewhere, we tried having just five performance buckets, which managers viewed 20 percent more favorably on some measures than the prior 41-category approach.

I can't underscore enough how hard this was for the People Operations team. Our work isn't life or death, but people screamed, people cried, people nearly quit. One of our challenges at Google is that because we afford Googlers so much freedom, because we are so data driven, and because Googlers care about fairness and how we treat one another, changes like this are Herculean efforts. Every team we approached was frustrated with the current system, and every team was resistant to doing something new. Within our YouTube division alone, there were a dozen different ideas about

---

[xliii] It was a great reminder of Alan Eustace's maxim that a great engineer is worth three hundred average ones, and that traditional performance and pay systems conspire to pay people based more on hierarchy than contribution.

what new rating systems to try. I couldn't be prouder of the perseverance, insight, and care with which the People Operations team worked through these changes, or more grateful to the teams that partnered with us to buck fifteen years of Google tradition to try something new.

Based on our experiments, in early 2013 we stopped doing quarterly ratings, in favor of every six months. There was some kvetching but no harm done. An instant time savings of 50 percent.

In late 2013, we moved more than 6,200 Googlers, representing almost 15 percent of the company, to a 5-point rating scale: needs improvement, consistently meets expectations, exceeds expectations, strongly exceeds expectations, and superb. Similar to the labels we had before, but with fewer discrete ratings.

We adhered to one of the core tenets of medicine: *Primum non nocere*. First, do no harm. Since this was our first rollout of this change, our goal was simply to achieve the same levels of satisfaction, fairness, and efficiency in the process as under our old rating scale. We figured that once we got through the initial skepticism and learning curve ("What do you mean I'm no longer a 3.8? I worked hard to be a 3.8!"), we'd save time by avoiding the agonizing over a tenth of a rating point, and managers would be forced to have more meaningful conversations with employees rather than hiding behind statements such as "Your score went up 0.1 this quarter. Great job and keep up the good work."

We were relieved to see that the loss of "precision" didn't hurt us. We compared how Googlers subject to the five-point scale felt relative to Googlers still on the 41-point scale. We asked:

- Were the right low performers identified?
- Were the right people identified for promotion?
- Were the discussions meaningful?
- Was the process fair?

Across the board, the new process was viewed as no worse than the old. While it seems a Pyrrhic victory, it was actually a huge relief for me. Some Googlers had worried that the loss of the precision conveyed by a 41-point rating scale would mean that our ratings would become less useful and meaningful. Instead, Googlers' survey responses revealed what we'd suspected all along: The forty-one points created only an illusion of precision.

Most Googlers admitted that for many ratings it wasn't possible to distinguish within plus or minus 0.1. For example, there was no consistent agreement about what a 3.1 was versus a 3.2. As Megan Huth, a member of our People and Innovation Lab, explained, "This created the possibility that the ratings were neither reliable nor valid. Given the same person and the same performance, she might be rated a 3.2 or a 3.3 depending on the rater and the calibration group. This means the rating isn't reliable. And if she gets a 3.3 when she's really a 3.2, then the rating also isn't valid—it doesn't reflect reality."

The ratings, therefore, really should have been, as Megan put it, "banded by error," meaning that we should have told people, "Jim, you're performing at a level somewhere within 3.3 to 3.5." But that's not what happened in practice. Managers would take the number and then ascribe real meaning to it, so if someone went from a 3.3 to a 3.5 it must be because they were improving, when in reality they could have been performing at the same level. And think of how much worse it would be if your rating dropped, and you were told it was because of your performance when in reality it was just measurement error.

And then something interesting happened. The 6,200 Googlers were spread across eight different teams at Google. But three teams, totaling over a thousand people, had decided to subdivide the five performance categories further. For example, one team added three subcategories to each category; so a star Googler could be rated "high superb," "medium superb," or "low superb." The next chart shows how the ratings ended up being distributed, though I've aggregated all the subcategories back into the five main ones so it's easy to see the difference between the two approaches. Group A stuck with five categories and Group B had fifteen.

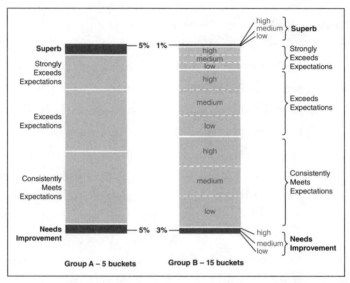

Average rating distribution of Groups A and B.

Group B, despite having more performance labels, which they hoped would create more differentiation across people, actually had far less differentiation than Group A. Five percent of Group A was "superb," but only 1 percent of Group B. I can tell you that all these teams in total performed at the same level. They added comparable value to Google and no team has better people than another. By simply having more rating categories to choose from, Group B unconsciously, inadvertently, and incorrectly decided that they have almost no star performers. Without meaning to, they dropped 80 percent of their top performers (four out of five) out of their top performance category.

By the time you're reading this, all of Google will have migrated to the 5-point rating scale. As of late 2013, it was still an experiment, but the early signs were good. First, it provided employees with more consequential feedback, replacing the murky differences between a 3.2 and a 3.3 rating. Second, it resulted in a wider performance distribution. As we shed performance rating categories, managers became more likely to use the extreme ends of the rating system. Despite the inconclusive academic research on performance rating systems and the neutral Googler feedback, we found that having five categories was superior to having more in at least these two ways.

By mid-2014, we saw even more positive results. We believed that different jobs have different opportunities for impact. If you're an engineer, your new product might benefit one hundred people or one billion. If you're a recruiter, try as you might, you simply don't have enough hours to impact one billion people. When we stopped offering guidance about what the right rating distribution ought to be, we saw four distinct rating patterns emerge that better reflected the actual performance characteristics of different teams and individuals.

We also saw that managers doubled their usage of the extremes of the rating system. Expanding the proportion of people receiving the top rating better reflected their actual performance (skip ahead to chapter 10 to see why that's true). And reducing the stigma of being in the bottom performance category made it easier for managers to have direct, compassionate conversations with their struggling employees about how to improve.

After much debate and consternation, we'd replaced a rating system that was imprecise and wasteful with a brand-new one that was simpler, more accurate, and required the same amount of time to calibrate ratings. To be clear, there's still ongoing debate and consternation! But we're working through it. And what we're already seeing is that as people become more comfortable with the new system, they appreciate it even more.

I share it with you here in the spirit of a beta launch, just as we release products when they're close enough to be far more useful than what's already out there, but before they're 100 percent polished and perfect.

That said, how many rating categories you have is the least important issue here, even though it was one that Googlers were stunningly passionate about. Don't offer fifteen-plus rating choices, but if you want to have three ratings or six, go ahead. I won't tell anyone.

Ensuring fairness

On the other hand, the soul of performance assessment is calibration. It's fair to say that without calibration, our rating process would be far less fair, trusted, and effective. I believe that calibration is the reason why Googlers were twice as favorable toward our rating system as people at other companies were to theirs.

So what is it? Google's rating system was (and is) distinctive in that it isn't just the direct manager making the decision. A manager assigns a draft rating to an employee—say, "exceeds expectations"—based on nailing OKRs but tempered by other activities, like the volume of interviews completed, or extenuating circumstances such as a shift in the economy that might have affected ad revenues.[xliv] Before this draft rating becomes final, groups of managers sit down together and review all of their employees' draft ratings together in a process we call calibration.

Calibration adds a step. But it is critical to ensure fairness. A manager's assessments are compared to those of managers leading similar teams, and they review their employees collectively: A group of five to ten managers meet and project on a wall their fifty to a thousand employees, discuss individuals, and agree on a fair rating. This allows us to remove the pressure managers may feel from employees to inflate ratings. It also ensures that the end results reflect a shared expectation of performance, since managers

---

[xliv] This is important. OKRs *influence* performance ratings but do not *determine* them.

often have different expectations for their people and interpret performance standards in their own idiosyncratic manner—just like in school, where some teachers were easy graders and others were tough. Calibration diminishes bias by forcing managers to justify their decisions to one another. It also increases perceptions of fairness among employees.[117]

The power of calibration in assessing people for ratings is not that different from the power of having people compare notes after interviewing candidates. The goal is the same: to remove sources of individual bias. Even if you're a small company, you'll have better results, and happier employees, if assessments are based on a group discussion rather than the whims of a single manager.

Even when calibrating, however, managers in group settings can make bad decisions. A host of errors in how we make decisions sneak in when we assess others. For example, recency bias is when you overweight a recent experience because it's fresh in your memory. If I had a great meeting with someone this week and then go into a calibration session where he is reviewed, I'm likely to inflate my estimate of him because I subconsciously lean on the recent, positive interaction. We addressed this by starting most calibration meetings with a single handout, describing the most common errors assessors make and how to fix them, a version of which is on the following page.

We'd start each calibration meeting by revisiting these errors. In the calibration sessions I attended, I observed that simply focusing managers on these phenomena, even for a moment, was enough to eliminate many of the distortions. Just as important, it created a language and cultural norm guarding against them. It's not uncommon today to hear someone in a calibration session redirect a conversation by saying, "Wait a minute. That's recency bias. We need to look at performance over the entire period, not just last week."

You can sense that even after reducing the frequency of our rating cycles and simplifying the scale we use to rate people, we still invest substantial time in the process. It may take just ten to thirty

Tips for Evidence-Based Calibration

Keep calibration data dri
7 common forms of cog

Cognitive Bias/ Group Dynamic	Definition	Example
Horns & Halo Effects	When the overall impression of someone as generally amazing/terrible clouds judgment against new evidence that might point to the contrary	"Tom is always such a ro᷍ some issues this quarter what a rock star he alwaᵧ
Recency Effect	Tendency to remember the last few things someone did and to weigh them disproportionately	"Tom is having a terrible two weeks he hasn't bee anything done."
Fundamental Attribution Error	Either paying too much attention to a person's "ability" and not enough to the situation/context that impacted their performance, or vice versa	"Tom bombed this proje᷍ he didn't get enough dir᷍ manager. He's great, I kn and he deserves a higheɾ "Tom bombed this proje᷍ me that he really can't g᷍ Where did we hire this g
Central Tendency	"Playing it safe" by rating close to the midpoint	"Well, 3.7 is a really high group, so what about mᴤ You're still giving the emᵖ 'exceeds expectations' m
Availability Bias	Mistaking what's easy to bring to mind with what's more frequent	"I remember Tom's first ᷍ thinking that I had neveɾ ramp up so fast. He's fan

Excerpt from a sample handout provided before performance calibration discussions. © Google, Inc.

minutes to assign draft ratings to your teams by ticking boxes in our performance management tool, but a calibration session can take three hours or even longer. Not every individual is discussed. Some time is spent making sure the calibrators are themselves calibrated, comparing individuals who are well known to more than one manager, so that they can then use that person as an anchor or benchmark. Calibrators also look at the distributions of ratings across different teams, not to force a single distribution but to understand why some teams might have different distributions. For example, one team might legitimately be stronger than another. Most of the time is then spent discussing cases that stand out for some reason, such as unusually fast performance acceleration or deceleration, large swings in performance, or borderline cases on the edge of rating categories.

When lots of other companies are abandoning ratings altogether, why do we stick with the system?

I think it's about fairness.

Ratings are tools, simplifying devices to help managers make decisions about pay and promotion. As an employee, I want to be treated fairly. I don't mind someone being paid more than me if they are contributing more. But if we're doing the same work and they're being paid way more, I'll be mighty unhappy. A just rating system means I don't have to worry about that. It also means that if someone does exceptional work, they'll be seen not just by their manager, but by lots of managers in the calibration meeting, who together create and promulgate a consistent standard across the company. Ratings also make it easier for people to move across the company. As a manager, I can trust that someone who "strongly exceeds expectations" does great work, whether her last job was working on Chrome or in Sales. As an employee, I can have confidence that people are being promoted based on merit, not politics. For a small team, you don't need this infrastructure—you know everyone. But once you have more than a few hundred people, employees are more comfortable trusting a reliable system than individual managers. Not because managers are necessarily bad or biased, but because a rating process that includes calibration actively weeds out badness and bias.

Avoid defensiveness and promote learning with one simple trick

A fair process for ratings gets you only so far. As a manager, you want to tell people not only how they did, but also how to do better in the future. The question is: What is the most effective way to deliver those two messages?

The answer: Do it in two distinct conversations.

Intrinsic motivation is the key to growth, but conventional performance management systems destroy that motivation. Almost everyone wants to improve. Traditional apprenticeship models are based on this notion. An inexperienced worker wants to learn, and will learn best when paired with a more expert partner who teaches them. Remember the first time you rode a bike, or learned to swim, or drove a car? The thrill of mastery, of accomplishment, is a powerful motivator.

But introduce extrinsic motivations, such as the promise of promotion or a raise, and the willingness and ability of the apprentice to learn starts to shut down. In 1969, Edward Deci and Richard Ryan of the University of Rochester escorted a series of individuals into a lab.[118] Each subject was given seven pieces of three-dimensional plastic that could be assembled into "millions of shapes." In each of three one-hour blocks of time, the individuals were told to reproduce four different shapes based on drawings they were given. If they couldn't solve one after thirteen minutes, the experimenter came in and helped them, to prove all the puzzles were solvable. To their right were drawings of other possible shapes, and to their left were the latest issues of *The New Yorker*, *Time*, and (this being the 1960's) *Playboy*. The experimenter sat in the lab with the subject, except for an eight-minute interlude in the middle of each hour-long block when he excused himself, ostensibly to score the results. He told subjects, "I shall be gone only a few minutes. You may do whatever you like while I'm gone." In reality, this was the key moment in the experiment. Left unsupervised, would the subject keep working on the puzzles?

Members of the control group spent about three and a half minutes (213 and 205 seconds) on the puzzle during the first two unsupervised windows, and four minutes (241 seconds) in the last window. Experimental-group members spent an average of four minutes (248 seconds) on the puzzle in the first window. Before the second hour started, experimental subjects were told they would receive a dollar for every puzzle they solved. With the added incentive, they spent more than five minutes (313 seconds) on the puzzle, 26 percent more time than in their first hour. Before the third hour, they were told that there was only enough money for one round of payment, so they wouldn't be paid any more. The time spent on the puzzle dropped to less than three and a half minutes (198 seconds), 20 percent less than in the first round and 37 percent less than in the paid round.

This was an early, small study, but it demonstrated the power of incentives, as well as the debilitating effect of removing the incentives. Deci and Ryan concluded that the introduction of an extrinsic reward caused people to think of their work differently from that point on by reducing intrinsic motivation. They went on to demonstrate that intrinsic motivation drives not just higher performance, but also better personal outcomes in terms of greater vitality, self-esteem, and well-being.[119] Workplaces that permit employees more freedom tap into that natural intrinsic motivation, which in turn helps employees feel even more autonomous and capable.

A similar dynamic exists when managers sit down to give employees their annual review and salary increase. The employees focus on the extrinsic reward—a raise, a higher rating—and learning shuts down. I once had a team member—I'll call him Sam—who would obsess each quarter about his rating. If it was higher, he didn't care why he earned the higher rating, or what behaviors he should be doing more of. If it was flat or lower, he would argue about why I didn't have all the facts and was wrong in my assessment. And Sam would keep arguing until I was so worn down I gave up and assigned him a higher rating. I'm ashamed to admit that, but know that I'm not alone.

In fact, employees have every reason to devote tremendous energy to arguing for higher ratings. As a manager, my incentive is to rate my employees fairly and honestly, so that the company's systems work. As an employee, my incentive is certainly to perform well, but it's also rational for me to argue, argue, argue with my manager to drive up my rating (as long as I don't push so hard my manager gets angry). It costs my manager nothing, save perhaps a little integrity (alas), and for me as an employee a higher rating means more money and opportunity. And I can afford to spend several hours a week preparing my arguments, while my manager not only doesn't have the time to do so for each of his employees, but will never have as much information as I do because he's not with me all day. As long as ratings are directly linked to pay and career opportunities, every employee has this incentive to exploit the system.

And even if I don't argue with my manager, he's worried that I might. In one study conducted by Maura Belliveau of Long Island University,[120] 184 managers were asked to allocate salary increases across a group of employees. The increases aligned nicely with performance ratings. Then they were told that the company's financial situation meant that funds were limited, but were given the same amount of funds to allocate. This time, men received 71 percent of the increase funds, compared to 29 percent for the women, even though the men and women had the same distribution of ratings. The managers—of both genders—had given more to the men because they assumed women would be mollified by the explanation of the company's performance, but that the men would not. They put more money toward the men to avoid what they feared would be a tough conversation.

We have an embarrassingly simple solution.

Never have the conversations at the same time. Annual reviews happen in November, and pay discussions happen a month later. Everyone at Google is eligible for stock grants, but those decisions are made a further six months down the line.

As Prasad Setty explains, "Traditional performance management systems make a big mistake. They combine two things that should be completely separate: performance evaluation and people development. Evaluation is necessary to distribute finite resources, like salary increases or bonus dollars. Development is just as necessary so people grow and improve."[121] If you want people to grow, don't have those two conversations at the same time. Make development a constant back-and-forth between you and your team members, rather than a year-end surprise.

The wisdom of crowds...it's not just for recruiting anymore!

We learned in chapter 5 that we make better hiring decisions about potential employees by relying on input from a crowd of people. The same principle holds for coaching and evaluating existing employees.[122] Going back to my team member Sam, I saw only a portion of his work, so he legitimately could argue that I didn't understand the full context of his performance. But Sam also had strong incentives to flatter me, represent his work in glowing terms, and denigrate the work of those around him so that he would look better in comparison. And he did all those things. It became almost impossible for me, as a manager, to have a perfect understanding of Sam's contribution.

His peers, however, saw the real Sam. They found him to be political, combative, and a bully. And I learned what they thought because once a year every Googler receives annual feedback not just from their manager, but also from their peers. When it's time to conduct annual reviews, Googlers and their managers select a list of peer reviewers that includes not just peers, but also people junior to them.

This feedback can be powerful. One leader, who had been cautious about wading into issues outside his area of expertise, was told, "Every time you open your mouth you add value." For years after, he told me, that one bit of insight from a colleague encouraged him to be a much more active member of his team. He'd been coached to speak up more by his boss, but it meant more coming from a peer.

In 2013, we also experimented with making our peer feedback templates more specific. Prior to that, we'd had the same format for many years: List three to five things the person does well; list three to five things they can do better. Now we asked for one single thing the person should do more of, and one thing they could do differently to have more impact. We reasoned that if people had just one thing to focus on, they'd be more likely to achieve genuine change than if they divided their efforts.

We used to ask individuals to list any and all accomplishments from the past year in a single, blank field. Now we asked them to list specific projects, their roles, and what they accomplished. We limited Googlers to 512 characters to describe what they did on each project,[123] figuring that if peer reviewers needed to read more explanation than that, they probably didn't know what the project was. And if they didn't already know the project, then peers were merely assessing the person's summary, not their actual work. Peer reviewers were then asked to rate (using a slider on the screen) how well they knew that particular project and how large the individual's impact was, and to add any comments. Over time, we get a sense of which feedback providers are reliable in their assessments, just as we do with interviewers. Googlers are also free to solicit feedback on specific topics at any point in the year, rather than waiting for a single day.

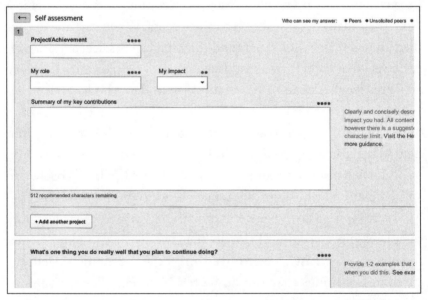

Excerpt from Google's self assessment templates. © Google, Inc.

To make sure employees' conversations with their managers were more useful, we developed a one-page handout for them to use during their performance conversation. Again, the goal was to make the conversation more specific and tangible. We distributed these handouts to employees just to be on the safe side; we hoped the managers would cover the right topics, but it didn't hurt to have the employees ready to guide the discussion too.

go/letstalkperf

Performance & development discussion guide for managers

This guide provides a framework to help you prepare and think through performance and development conversations with your team. You can use this guide whether you're holding a full review (e.g., discussing peer feedback and your written manager assessment) or a mid-year check in (e.g., sharing the most recent rating).

Development conversations as part of the official Perf review cycles are just one opportunity for you to connect with your Googlers. Sharing feedback and discussing how they can grow is an ongoing part of your role as a manager. You can also use this framework to structure performance and development conversations that you hold throughout the year, building upon past discussions.

Key areas to cover:

    Getting started
    1. Overall performance
    2. What to keep doing & next steps
    3. What to improve on & next steps
    4. [optional] Longer-term goals
    5. Recap

Additional resources:
- *You may find it helpful to leverage this tracking sheet as you compile information for each individual, and/or this worksheet to share directly with your Googler*
- *We have also shared this conversation guide with Googlers to help them prepare for these discussions*

Getting started

Before you dive in, ensure the goals of the conversation are clear - are you discussing a full review incl. peer feedback, are you discussing the last 6 months and the related perf rating, or are you checking in mid cycle?

What to cover:	Things to consider:
• Articulate the goal and structure of the conversation • Have examples ready to enrich the discussion • Ask questions and encourage your Googler to speak openly	• Past development conversations with your Googler • How does your Googler best receive and integrate feedback? If you feel unsure, this could be something to discuss • Think about and combat any potential biases - the checklists at go/bbPerf will help

Excerpt from Google's discussion guides for managers. © Google, Inc.

I was surprised, maybe even a little embarrassed (I'm supposed to know this stuff!), at how a few small changes could have such large effects across the board. Making the templates more specific reduced the time spent writing reviews by 27 percent, and for the first time, 75 percent of peers felt that writing the reviews was helpful, up 26 points (on a scale of 100) from the prior year. Those using the discussion guide with their managers rated their performance conversations 14 points more favorably than those who didn't. As one effusive Googler wrote, "OMG this version is so much easier and takes SO much less time. Thank you for giving me my September back!!!"

The experiments gave us the confidence, and credibility with Googlers, to roll out these changes to the entire company in 2014. And Googlers have been happier. Eighty percent of all Googlers now agreed that providing feedback this way was time well spent, up from 50 percent two years earlier. Still not perfect, but massive improvements.

Putting it all together for promotions

In most companies, if your ratings are high enough, you get promoted. Often your boss makes the call, or you transfer into a new job and get a fancy new title. At Google it doesn't quite work that way. By now you'll guess that promotion decisions, like rating decisions, are made by committees. They review people who are up for promotion and calibrate them against promoted people from prior years and well-defined standards, to ensure fairness.

And it wouldn't be Google if we didn't also rely on the wisdom of crowds. Peer feedback is an essential part of the technical promotion packet that committees review.

There's just one other twist. Googlers working in engineering or product management can nominate themselves for promotion.[xlv] Interestingly enough, we found that women are less likely

---

[xlv] With the caveat that some of my best friends are salespeople, I'll point out that salespeople are far more promotion-seeking than engineers. As bizarre as it may sound to non-engineers, most of our engineers aren't motivated at all by level and status. They just want to work on cool stuff. They don't obsess about the next rung of the ladder the way nontechnical people can. When I last discussed self-nomination with some of our sales leaders, they had some trepidation about the floodgates of self-nominations this approach might unleash. My counterargument was that after

to nominate themselves for promotion, but that when they do, they are promoted at slightly higher rates than men. This seems to be related to the dynamic that is seen in classrooms: In general, boys raise their hands and try to answer any question. Girls tend to wait to be certain, even though they are right as often as boys, if not more often.[124]

We also found that with a small nudge (an email from Alan Eustace to all technical Googlers describing this finding), women then nominate themselves in the same proportion as men. His latest note shared our promotion statistics by gender and level with everyone, and my favorite lines were these:

> I wanted to update everyone on our efforts to encourage women to self-nominate for promotion. This is an important issue, and something I feel passionately about. Any Googler who is ready for promotion should feel encouraged to self-nominate and managers play an important role in ensuring that they feel empowered to do so....We know that small biases—about ourselves and others—add up over time and overcoming them takes conscious effort....To monitor this, we also reviewed the last three cycles of promotion data to identify any persistent gaps....I will continue to share this data in an effort to be transparent and open about the issue and so we can keep up this positive momentum.

Of course, not everyone gets promoted, regardless of gender. If you are not promoted, the committee provides feedback on what to do to improve your chances next time. Seems obvious when you read it, but it's a vanishingly rare practice. As you can imagine, at our size it takes hundreds of engineers to populate these committees, and

---

a couple of cycles of high-quality feedback to people about why they are not getting promoted, the system would settle down and work. I've not yet won this argument, but continue fighting!

the promotion process can easily take two or three days each cycle. We find that, perhaps because of the populist committee structure, the time spent on it, and the fact that committee members have no incentive other than to make good decisions (just like our hiring committees), engineers are more likely to rate their promotion process as fair than non-engineers.

A new hope

The only companies I've seen that spend as much time on performance management and promotions as we do are colleges and firms run like partnerships. At both, promotion eventually leads to you becoming something like a permanent part of the family, either as a tenured professor or a partner. Great care is taken in making a long-term commitment to you.

We put the same care into our reviews out of necessity. Google's revenues and headcount have grown roughly 20 to 30 percent in each of the past five years. We do our best to hire people who have a proven aptitude for learning, and then do everything we can to help them grow as fast as they can. Making sure our people are developing is not a luxury. It's essential for our survival. But the fundamental concepts we've had to evolve make up a language that translates to just about any company.

First, set goals correctly. Make them public. Make them ambitious.

Second, gather peer feedback. There is a range of online tools, not the least of which is Google Sheets, that allows you to create surveys and compile the results. (Type "Google Spreadsheets survey form" into your browser.) People don't like being labeled, unless they are labeled as extraordinary. But they love useful information that helps them do their jobs better. It's this latter piece that most companies miss. Every company has some kind of evaluation system that is then used to distribute rewards. Few have equally disciplined mechanisms for development.

Third, for evaluation, adopt some kind of calibration process.

We prefer meetings where managers sit together and review people as a group. It takes more time, but it gives you a reliable, just process for assessment and decision-making. A side effect is that it's good for the culture for people to sit together, reconnect, and affirm what we value. In-person meetings are most efficient for companies with up to ten thousand people. After that, you need an awful lot of conference rooms to fit everyone. We keep at it and make it work for more than fifty thousand people because it serves our people well.

Fourth, split reward conversations from development conversations. Combining the two kills learning. This holds true at companies of any size.

All the other pieces of performance management: the number of performance categories, whether the categories are numbers or words, how often to give ratings, whether to do it online or on paper...it doesn't matter. After a long time in the desert, we found a set of ratings and a rhythm that work for us, but there's no external evidence either way on any of this stuff. So unless you want to run the same experiments we did in search of different results, I wouldn't worry about these other bits.

Focus instead on what does matter: a fair calibration of performance against goals, and earnest coaching on how to improve. The Lisa Simpson in all of us wants to be evaluated because she wants to be the best. She wants to grow. All you have to do is tell her how.

....

WORK RULES...FOR PERFORMANCE MANAGEMENT

- ☐ Set goals correctly.
- ☐ Gather peer feedback.
- ☐ Use a calibration process to finalize ratings.
- ☐ Split rewards conversations from development conversations.

....

8

························

The Two Tails

The biggest opportunities lie in your absolute
worst and best employees

Your team has tails.

Anything you can measure follows some kind of distribution from low to high, small to big, near to far. Remember when you were a kid and your teacher would have the class line up from shortest to tallest? I was always on the taller end and it was easy to sort us. There were maybe three or four of us in a class of thirty who would know to head to the right end of the line, and a handful of kids who were the smallest who would head to the left. Then there would be twenty or more kids, all within an inch or two of the same height, left milling around in a clump as they haltingly sorted themselves into a line.

It turns out teachers have enjoyed sorting students into groups by height for at least a century.

In 1914, Albert Blakeslee, of what is now known as Connecticut College, asked his students to arrange themselves into lines based on height. Just like in your own classroom, most of the students end up clumped in the middle, with some out at the extreme edges. The heights of the college students follow a distribution, ranging from 4 feet 10 inches to 6 feet 2, and you can clearly see the concentration in the center.

Number of individuals in each rank 1   0   0   1   5   7   7   22   25   26   27   17   11   17   4   4   1
Heights in feet and inches to which
ranks correspond 4:10 4:11 5.0 5:1 5:2 5:3 5:4 5:5 5:6 5:7 5:8 5:9 5:10 5:11 6:0 6:1 6:2

Living histogram of 175 male college students.[125]

The "tails" of the distribution are the team members at the extremes, say below 5 feet 4 inches tall and above 5 feet 11 inches. They are the bottom and top 10 percent of the distribution in the example below.

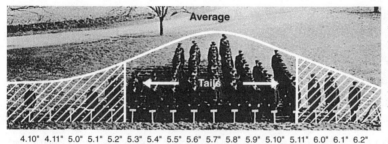

4.10" 4.11" 5.0" 5.1" 5.2" 5.3" 5.4" 5.5" 5.6" 5.7" 5.8" 5.9" 5.10" 5.11" 6.0" 6.1" 6.2"

The heights of the students follow a normal distribution, with each of the two "tails" representing those with "extreme" heights.

The same holds true for Googlers—we asked Googlers to line up by height and the result was a normal distribution with the same two tails.[xlvi]

[xlvi] Googlers had more fun with this task than I thought they would.[126]

A distribution describes the pattern made by data. Height happens to be best described by a kind of distribution called "normal." It's also known as the bell curve due to its shape and as a Gaussian distribution, after Carl Friedrich Gauss, who described it in a paper in 1809.[127]

The Gaussian distribution is popular with researchers and businesspeople because it describes the distribution of many things: height, weight, extroversion and introversion, the width of tree trunks, the size of snowflakes, the speed of cars on a highway, the rate at which defective parts are made, the amount of customer service calls that come in, and so on. Even better, anything that follows a Gaussian distribution will have an average and a standard deviation, and you can use those to predict the future. The standard deviation describes how likely a certain amount of variation (or deviation) is to happen. For example, the average woman in the United States is 5 feet 4 inches tall,[128] with a standard deviation of just under three inches. This means that 68 percent of women are between 5 feet 1 inch and 5 feet 7 inches. That's one standard deviation. Ninety-five percent are within two standard deviations of the average, between 4 feet 10 inches and 5 feet 10 inches. And 99.7 percent are no more than three standard deviations away from the average, between 4 feet 6 inches and 6 feet 2 inches. If you look around your office or neighborhood, you'll see this feels about right. (For men, the average height is 5 feet 10 inches tall, with the same range of about plus or minus 3 inches at every standard deviation. You might notice that in Blakeslee's photo the average man is about 5 feet 7 inches. Thanks to better nutrition, Americans have grown taller over the last hundred years.)

The virtue of the Gaussian distribution is also its weakness. It's so easy to use, and superficially it seems to describe so many different phenomena that it's applied in cases where it doesn't describe the underlying reality. A Gaussian distribution will massively underpredict the frequency of major physical and economic events (enormous earthquakes, hurricanes, and stock market swings), the dramatic span in economic outcomes for people (the expanding gap between the poor and the top 1 percent), and the exceptional human performance of a small number of individuals (Michael Jordan compared to other basketball players of his day). The 2011 Japan earthquake (magnitude 9.0), Bill Gates's net worth (over $70 billion), and even the population of New York City (8.3 million people) are too far from average to show up as a likely scenario in a Gaussian model, yet we know they exist.[129]

Statistically, these phenomena are better described by a "power law" distribution, which is compared to a Gaussian distribution below.

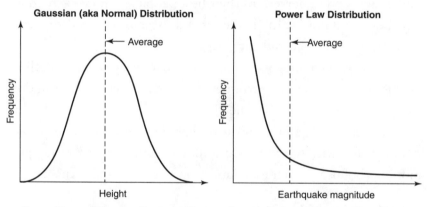

Comparison of the distribution of human height and earthquake magnitude. Height varies evenly around an average with roughly half of people above and half below average in height. In contrast, the large majority of earthquakes are below average size.

The name "power law" is used because if you wrote an equation describing the shape of the curve, you'd need to use an exponent to describe it, where one number is raised to the power of another number (e.g., in $y = x^{-1/2}$, the exponent is $-\frac{1}{2}$ and x is "raised to the power of $-\frac{1}{2}$." This equation gives you a graph roughly like what's shown on the top right).

Most companies manage people using a normal distribution, with most people labeled as average and two tails of weak and strong performers pushed out to the sides. The tails aren't as symmetrical as when you look at height, because failing employees get fired and the worst don't even make it in the door, so the left tail is cut short. Companies also treat people as if their actual output follows the same distribution. That's an error.

In fact, human performance in organizations follows a power law distribution for most jobs. Herman Aguinis and Ernest O'Boyle of Indiana University and the University of Iowa explain that "instead of a massive group of average performers dominating…through sheer numbers, a small group of elite performers [dominate] through massive performance."[130] Most organizations undervalue and underreward their best people, without even knowing they are doing it. In chapter 10 I'll explain why and suggest a better way to manage and pay people.

For now, it's enough to appreciate that every team has tails—people who sit at either extreme of the performance distribution. Most companies get rid of the "bottom tail" performers, who live in a purgatory of failure and fear that they could be fired at any moment. For the "top tail," life couldn't be better, with ready promotions, bonuses, and the adulation of their peers and management.

Help those in need

What most organizations miss is that people in the bottom tail represent the biggest opportunity to improve performance in your company, and the top tail will teach you exactly how to realize that opportunity.

Jack Welch popularized the "up or out" model of management described in the first pages of this book, where General Electric employees were rated each year, and the bottom 10 percent were fired: You either moved up in the organization or you were moved out.

But isn't there a cost to this? It takes time and money to recruit new people, who often are more expensive than current employees and who need to learn the new jobs—and even then may not succeed! In a study of over a thousand research analysts at investment banks, Professor Boris Groysberg of Harvard Business School found that "star analysts suffer an immediate and lasting decline in performance" when they move to a new company.[131] Their prior success had been dependent on their coworkers, resources available to them, their fit with the company's culture, and even the personal reputation or brand they had built up.

In an ideal world, you would have hired all the right people to begin with, and if you're running an objective, well-calibrated hiring process you're probably pretty close. But even then, you will make mistakes, and those people will sink to the low end of your performance curve.

At Google, we regularly identify the bottom-performing 5 percent or so of our employees. These individuals form the bottom tail of our performance distribution. Note that this happens outside our formal performance management process. We're not looking to fire people: We're finding the people who need help.

I confess that we don't have a reliable absolute measure for performance for every job, and we don't force a distribution of ratings, because different teams perform at different levels. It would be madness to force the manager of a team of all superstars to rank someone as failing. So this is a human process, not an algorithmic one, where managers and the People Operations team look at individuals. In practice, the bottom tail does end up including those who "need improvement," but it also captures "skimmers," people who have been skimming along the lower end of meeting expectations for a long time. Because we track the 5 percent only at the highest organizational levels, some groups end up with no one in this group and some end up with more than 5 percent. We wondered if we should be firing these performers, as many other companies do, but that would have meant culling 20 percent of our employees each year (5 percent each quarter). It would also have implied that our hiring approach wasn't working. If we were successfully screening for people who are more than "brains on a stick"—brilliant, adaptable, conscientious Googlers—we shouldn't need to conduct regular cullings.

So rather than following the traditional path of making "poor performance" the kiss of death, we decided to take a different approach: Our goal is to tell every person in the bottom 5 percent that they are in that group. That is not a fun conversation to have. But it's made easier by the message we give these people: "You are in the bottom 5 percent of performers across all of Google. I know that doesn't feel good. The reason I'm telling you this is that I want to help you grow and get better."

In other words, this isn't a "shape up or ship out" conversation; it's a sensitive talk about how to help someone develop. A colleague once described it as "compassionate pragmatism." Poor performance is rarely because the person is incompetent or a bad person. It's typically a result of a gap in skill (which is either fixable or not) or will (where the person is not motivated to do the work). In the latter case, it could be a personal issue or a useful sign that there is something bigger wrong with the team that needs to be addressed.

In fact, the way we de-emphasize role-related knowledge in hiring leaves us a bit vulnerable to this issue, because we like to hire people who may not know how to do a job. We have faith that almost all of them will figure it out, and along the way are more likely to invent a novel solution than someone who's "been there, done that."

When they don't, we first offer a range of training and coaching to help them build their capabilities. Note that this is very different from the typical approach of hiring people and then trying to train them into being stars. Our interventions here are for the small handful of people who struggle most, rather than for everyone. If that doesn't work, we then help the person find another role within Google. Typically, this results in the person's performance improving to average levels. This may not sound like much, but think about it this way: Out of a group of a hundred people, Jim was one of the five worst performers. After this intervention, Jim was about the fiftieth-best performer.[132] Not a superstar, but Jim is now contributing more than forty-nine other people, where before he had been better than only three or four people. What would your company be like if all the worst people got that much better? And if even the bottom forty-nine were still better than the competition?

For the remaining people, some choose to leave and others we need to fire. It sounds harsh, but they tend to end up happier because we've shown sensitivity to their situation and invested in them along the way, and we give them time to find an organization where they can excel. I once had to terminate someone who worked for me, who upon exiting told me, "I'd never be able to do your job." I said, "You can, but at a place where the demands are different." Three years later, he called me to share that he'd been promoted to chief human resources officer of a Fortune 500 company, and was thriving. He said the pace was a bit slower than Google, but it fit him perfectly. And he'd been able to become a trusted counselor to the CEO precisely because of his measured, thoughtful style.

This cycle of investing in the bottom tail of the distribution means your teams improve... a lot. People either improve dramatically or they leave and succeed elsewhere.

What's illuminating is that even "Neutron Jack," as Welch was called due to the layoffs and firings that characterized much of his tenure as CEO, mellowed in his description of the process in later years. In 2006, he elaborated on his approach to "rank-and-yank" sorting of employees:

> The "yank" myth of differentiation says the bottom 10% are summarily fired. In reality, that's rare. More typically, when a person has been in the bottom 10% for a sustained period of time, the manager starts a conversation about moving on. Occasionally, of course, an underperformer doesn't want to go. But confronted with the cold reality of

how the organization views them, most people leave of their own accord and very often end up at companies where their skills are a better fit and they are more appreciated.[133]

He then argues that it's kinder to be direct with an employee:

Compare that with companies where managers, in the name of kindness, allow people, and particularly underperformers, to plod along for years. Then a downturn occurs. Middle-aged underperformers are always the first to get the ax. One by one, their manager calls them in for a conversation that usually goes like this:
"Joe, I'm afraid you have to leave."
"What! Why me?"
"Well...you were never very good."
"I've been here 20 years. Why didn't you ever tell me?"
Why not indeed? The employee, years earlier, might have been able to find a job with a future. Now, at age 45 or 50, he must enter a job market more competitive than ever. That's cruel.[134]

I can't underscore enough that our identifying the bottom 5 percent is *not* the same as "stack ranking" employees, a method of forcing all employees to fall into performance categories with a fixed distribution. That approach can poison a culture as employees turn on one another in a fierce struggle to avoid being at the bottom. Kurt Eichenwald wrote a scathing indictment of stack ranking in *Vanity Fair* in 2012:[135]

Every current and former Microsoft employee I interviewed— *every one*—cited stack ranking as the most destructive process inside of Microsoft, something that drove out untold numbers of employees.... "If you were on a team of 10 people, you walked in the first day knowing that, no matter how good everyone was, two people were going to get a great review, seven were going to get mediocre reviews, and one was going to get a terrible review," says a former software developer. "It leads to employees focusing on competing with each other rather than competing with other companies."

Almost exactly a year later, in November 2013, Microsoft's head of HR, Lisa Brummel, emailed employees announcing the abolishment not just of stack ranking, but of *all* ratings.[136]

As I wrote in chapter 2, if you believe people are fundamentally good and worthy of trust, you must be honest and transparent with them. That includes telling them when they are lagging behind in their performance. But having a mission-driven, purposeful workplace also requires that you approach people with sensitivity. Most people who are performing poorly know it and want to get better. It's important to give them that chance.

Put your best people under a microscope

At the same time, the top tail, the very best performers, experience a company differently than average or mediocre performers do. Our data show us that they find it easier to get things done, feel more valued, feel that their work is more meaningful, and leave the company at one-fifth the rate that our lowest performers do. Why? Because top performers live in a virtuous cycle of great output, great feedback, more great output, and more great feedback. They get so much love on a daily basis that the extra programs you might offer can't actually make them much happier.

More important is to learn from your best performers.[xlvii]

[xlvii] It's also important to compare your best performers to your worst performers. As Kathryn Dekas of our People and Innovation Lab explains, "If you only study the people you're trying to emulate, you might conclude that the key behaviors contributing to their success are the behaviors common among most or all of them.

Every company has the seeds of its future success in its best people, yet most fail to study them closely. This is a missed opportunity, because as Groysberg demonstrates, high performance is highly context dependent. Benchmarking and best practices tell you what worked elsewhere, but not what will work for you.

In contrast, understanding precisely what makes your best people succeed in your unique environment is the natural extension of Groysberg's findings. If success depends on specific, local conditions, then you are best served by studying the interplay of high performance and those local conditions.

As you might expect, we study our best people very closely at Google. In 2008, Jennifer Kurkoski and Brian Welle cofounded the People and Innovation Lab (PiLab), an internal research team and think tank with the mandate to advance the science of how people experience work. Many PiLab scientists have PhDs in psychology, sociology, organizational behavior, or economics, and have moved into leadership roles that enable them to apply their research skills to tricky organizational questions and challenges. Neal Patel and Michelle Donovan are typical: Michelle has been instrumental in changing our approach to performance management, Neal leads the Human/Social Dynamics program in Google's Advanced Technology and Projects group. Their initial research agenda gives a sense of what you can learn by studying your best people:

- Project Oxygen initially set out to prove that managers don't matter and ended up demonstrating that good managers were crucial.

This seems like a reasonable conclusion. But it's also possible that the worst performers have the same behaviors, and you'll never know unless you examine them as well. You can easily end up identifying the wrong behaviors as leading to success if you don't also study the other groups....In technical terms this is called 'sampling on the dependent variable.'" This is yet another variant of the sample bias we discussed in chapter 6, and why "best practices" can often be misleading.

- Project Gifted Youngsters was targeted at explaining what people who sustain the highest performance for long periods of time do differently from everyone else. They explored the top 4 percent compared to the other 96 percent, then dug into the top 0.5 percent versus the other 99.5 percent.

- The Honeydew Enterprise (named for Bunsen Honeydew, the intrepid innovator from the Muppets) strove to understand the behaviors and practices that most foster and inhibit innovation among software engineers.

- Project Milgram explored the most effective ways to mine social networks for knowledge within Google. (This was named for the same Stanley Milgram who studied obedience. As Jennifer Kurkoski told me: "He conducted the initial small-world experiment, in which randomly chosen individuals in Omaha or Wichita were asked to start a chain of mail with the intended end point being a named individual in Boston. The average number of 'hops' for these chains of mail was 5.5, leading to the popular acceptance of the concept of six degrees of separation.")

Project Oxygen has had the most profound impact on Google. The name comes from a question Michelle once asked: "What if everyone at Google had an amazing manager? Not a fine one or a good one, but one that really understood them and made them excited to come to work each day. What would Google feel like then?" Neal was in the habit of naming his projects after elements from the periodic table, so Michelle proposed Project Oxygen, because "having a good manager is essential, like breathing. And if we make managers better, it would be like a breath of fresh air."

So what was Project Oxygen looking to accomplish? The hypothesis was that manager quality had no impact on team performance. Neal explained: "We knew the team had to be careful. Google has high standards of proof, even for what, at other places, might be considered obvious truths. Simple correlations weren't going to be enough. So we actually ended up trying to prove the opposite case—that managers *don't* matter. Luckily, we failed."[137]

Engineers at Google deeply believed that managers don't matter. On the face of it, that may sound preposterous. But you have to understand how much engineers hate management. They don't like managers and they certainly don't want to become managers.

Engineers generally think managers are at best a necessary evil, but mainly they get in the way, create bureaucracy, and screw things up. It was such a deeply held belief that in 2002 Larry and Sergey eliminated all manager roles in the company.

We had over three hundred engineers at the time, and anyone who was a manager was relieved of management responsibilities. Instead, every engineer in the company reported to Wayne Rosing. It was a short-lived experiment. Wayne was besieged with requests for expense report approvals and for help in resolving interpersonal conflicts, and within six weeks the managers were reinstated.[138]

Managers clearly served some purpose, but by 2009 engineers' innate leeriness about management had reasserted itself. In the intervening seven years we'd added over nineteen thousand employees, most of whom came from traditional environments where managers were largely unhelpful, if not downright destructive. We saw this problem in our recruiting as well, particularly outside the United States. Our hiring credo was that an engineering manager had to be at least as technically capable as her team.[xlviii] When that wasn't the case, the manager wasn't respected and was known as a NOOP, a term borrowed from computer science that means "no operation performed." While in the United States there was some history of companies having parallel tracks for technical individual contributors and for managers (IBM, for example, pioneered

---

[xlviii] Even though we like hiring smart generalists, there are some fields—like engineering, tax, or law—where some baseline level of expertise is essential.

an individual-contributor promotion track where you could receive the same rewards and titles as a manager purely on the basis of your technical achievements), in Asia and Western Europe it was far more common for engineers to be promoted to management roles and then remove themselves from day-to-day engineering. As a result, we often rejected senior candidates who might have been good managers but were too distant from technical issues.

Everyone has an idea of what a good or bad manager is, but it's a subjective standard. Michelle and Neal wanted to be consistent in how they framed their comparison, so they relied on two quantitative sources of data: performance ratings and Googlegeist results. They calculated the average performance rating for each manager, looking back at the last three performance periods. As a proxy for how teams would rate their managers' quality, they analyzed each manager's Googlegeist results, which asked everyone in the company what they thought of their manager's performance, conduct, and support. They sorted our managers into four quadrants:

How managers were initially sorted for Project Oxygen.

The key was to really understand the best of the best and the worst of the worst. What were those managers doing to get such different results? To find the answer, they looked for the absolute extremes of performance. Out of more than a thousand managers, only 140 scored in the top 25 percent as individual performers and in Googlegeist. Even fewer, 67, were in the bottom 25 percent for both measures. This at least was an encouraging sign: We had twice as many managers in the "best of the best" category as in the "worst of the worst."

Subsequent refinement to identify managers who were in the top or bottom 25 percent in both team happiness and performance.

To be in the top 25 percent, your team needed to be only 86 percent positive about you, barely above the average of 84 percent. And the cutoff to be in the bottom quartile was 78 percent, still not far off average. At this point, it looked like the engineers were going to be proven right. There didn't seem to be a big difference between the best and worst managers.

So Michelle and Neal looked deeper. When they took apart the factors that added up to the overall manager-happiness score, they found some very big differences. Googlers with the best managers did 5 to 18 percent better on a dozen Googlegeist dimensions when compared to those managed by the worst manager. Among other things, they were significantly more certain that

- Career decisions were made fairly. Performance was fairly assessed and promotions were well deserved.
- Their personal career objectives could be met, and their manager was a helpful advocate and counselor.
- Work happened efficiently. Decisions were made quickly, resources were allocated well, and diverse perspectives were considered.
- Team members treated each other nonhierarchically and with respect, relied on data rather than politics to make decisions, and were transparent about their work and beliefs.
- They were appropriately involved in decision-making and empowered to get things done.
- They had the freedom to manage the balance between work and their personal lives.

Teams working for the best managers also performed better and had lower turnover. In fact, manager quality was the single best predictor of whether employees would stay or leave, supporting the adage that people don't quit companies, they quit bad managers.

But, some argued, there were only 207 "best" and "worst" managers across the whole company, a pretty small sample. How could we be sure the differences were because of the managers? Maybe some managers just ended up with stronger, happier teams by chance. The only way to test if the performance and happiness differences were really due to the managers would be to shuffle people randomly across teams and hold everything else constant, to see if changing only the manager made a difference. Even Google wouldn't be crazy enough to randomly mix up teams and managers just for the sake of knowledge, would we?

Fortunately, we didn't need to. Googlers ran the experiment for us by switching teams. Engineers are free to change project teams throughout the year, but had no way of knowing whether a prospective manager was among the "best" or "worst." In 2008, sixty-five Googlers moved from "best" managers to "worst" managers, and sixty-nine moved in the other direction. All of these were typical Googlers, performing well and satisfied with the company.

And did managers ever make a difference! The sixty-five people who moved to worse managers scored significantly lower on thirty-four of forty-two Googlegeist items. The next year, those moving to better managers saw significant improvements on six of the forty-two items. And the biggest changes were on questions that measured retention, trust in performance management, and career development. Switching to a worse manager was—by itself—enough to transform someone's experience of Google, chipping away at their trust in the company and causing them to consider quitting.

So managers did matter. And not only that, but amazing managers mattered a lot. We now knew who our best and worst managers were, but we didn't know what they did differently. Our analysis was descriptive but not prescriptive. How could we figure out what the best managers were doing differently from the worst, and how could we then turn that into an engine that would continuously improve the quality of managers at Google?

Proving that not all research requires teams of brainiac researchers, we took a very simple approach to finding out what the best and worst managers did differently: We asked them. A sample of managers was interviewed by Googlers who had been given interview guides to follow, but who didn't know whether the managers they were interviewing were good, bad, or mediocre. This is called a double-blind interview methodology, because it prevents the interviewer from biasing the interviewee, and the interviewee doesn't know which category they are in either. In other words, both the interviewer and interviewee are "blind" to the experimental condition. Michelle and Neal validated those findings by comparing them to the endorsements written for our Great Manager Award (a program where the best twenty managers are selected based on Googlers' write-in nominations), to the Googlegeist comments employees made about their managers, and to peer feedback written for managers. They were checking to see if the behaviors the managers claimed were the reasons for their success or struggles were also the behaviors that affected Googlers.

The research showed eight common attributes shared by high-scoring managers and not exhibited by low-scoring managers:

The 8 Project Oxygen Attributes

1. Be a good coach.
2. Empower the team and do not micromanage.
3. Express interest/concern for team members' success and personal well-being.
4. Be very productive/results-oriented.
5. Be a good communicator—listen and share information.
6. Help the team with career development.
7. Have a clear vision/strategy for the team.
8. Have important technical skills that help advise the team.

List of the 8 Project Oxygen attributes from Google. © Google, Inc.

We now had a prescription for building great managers, but it was a list of, quite frankly, pretty dull, noncontroversial statements. To make it meaningful and, more important, something that would improve the performance of the company, we had to be more specific. For example, of course the best managers are good coaches! Superficially this seems obvious, but most managers, if they have regular one-on-one meetings at all, just show up and ask "What's going on with you this week?" Most don't hold regular 1:1 meetings where they partner with the employee to diagnose problems and together come up with ideas tailored to the employee's strengths. Most don't combine praise and areas to work on. The specific prescription for managers is to prepare for meetings by thinking hard about employees' individual strengths and the unique circumstances they face, and then use the meeting to ask questions rather than dictate answers. And unexpectedly, we found that technical expertise was actually the least important of the eight behaviors across great managers. Make no mistake, it is essential. An engineering manager who can't code is not going to be able to lead a team at Google. But of the behaviors that differentiated the very best, technical input made the smallest difference to teams.

In addition to being specific, we had to make good management automatic. Atul Gawande has written persuasively in *The New Yorker* and in his book *The Checklist Manifesto* about the power of checklists. I first encountered his writing in the 2009 article "The Checklist,"[139] where he described the test flight of the Model 299, a next-generation long-range bomber developed by the Boeing Corporation in 1935. It could "carry five times as many bombs as the Army had requested... fly faster than previous bombers, and almost twice as far." The only problem was that it crashed.

The plane was more complex than its competitors, and on its maiden voyage the experienced pilot had "forgotten to release a new locking mechanism on the elevator and rudder controls," resulting in the deaths of two of the five crew. The Army's solution wasn't more training. It was a checklist. Gawande concludes: "With the checklist in hand, the pilots went on to fly the Model 299 a total of 1.8 million miles without one accident... dubbed [it] the B-17... [and] gained a decisive air advantage in the Second World War which enabled [the] devastating bombing campaign across Nazi Germany." Gawande then argued that the field of medicine was entering the same stage, where complexity outstripped human capacity and checklists would save lives.

Reading this, I realized that management too is phenomenally complex. It's a lot to ask of any leader to be a product visionary or a financial genius or a marketing wizard as well as an inspiring manager. But if we could reduce good management to a checklist, we wouldn't need to invest millions of dollars in training, or try to convince people why one style of leadership is better than another. We wouldn't have to change who they were. We could just change how they behave.

So Michelle, Neal, and a growing team of People Operations members created a system of reinforcing signals to improve the quality of management at Google. The most visible, a semiannual Upward Feedback Survey, asks teams to give anonymous feedback on their managers:

Sample UFS Feedback Questionnaire

1. My manager gives me actionable feedback that helps me improve my performance.
2. My manager does not "micromanage" (i.e., get involved in details that should be handled at other levels).
3. My manager shows consideration for me as a person.
4. My manager keeps the team focused on our priority results/ deliverables.
5. My manager regularly shares relevant information from his/her manager and senior leadership.
6. My manager has had a meaningful discussion with me about my career development in the past six months.
7. My manager communicates clear goals for our team.
8. My manager has the technical expertise (e.g., coding in Tech, accounting in Finance) required to effectively manage me.
9. I would recommend my manager to other Googlers.

Sample Upward Feedback Survey manager questionnaire from Google. © Google, Inc.

The survey itself is the checklist. If you perform every behavior on the list, you'll be an amazing manager.

Results are reported to each manager in the following way:

UFS Report for Craig Rubens

Overall Percent Favorable:	91% (?)
Top Quartile Overall:	93%
Bottom Quartile Overall:	75%

Fav	Neutral	Unfav

■ **% Favorable** - the percent of Googlers who selected "agree"/"strongly agree" to the given item

☐ **% Neutral** - the percent of Googlers who selected "neutral" to the given item

■ **% Unfavorable** - the percent of Googlers who selected "disagree"/"strongly disagree" to the given item

Detailed Results

Here are the survey results for Googlers who reported directly to you as of January 1, 2015. We display the current items where three or more people responded.

Filter: – All Oxygen Attributes – ⬦ *Hover over any item to view its Oxygen Attribute (learn more)

	Item	N	% Favorable	Vs Prior Fav	Vs Global Business Fav	Find resources
1	My manager does not "micromanage" (i.e., get involved in details that should be handled at other levels).	6	100	0 Vs Q1-2013 Googlegeist / 0 Vs Q3-2012 UFS	+17	📚
2	My manager balances giving freedom with being available for advice.	6	100	+20 Vs Q3-2012 UFS	+12	📚
3	My manager makes it clear he/she trusts the team.	6	100	0 Vs Q3-2012 UFS	+15	📚
4	My manager shows consideration for me as a person.	6	100	+14 Vs Q1-2013 Googlegeist / 0	+9	📚

Sample Upward Feedback Survey manager report from Google.
Data is illustrative. © Google, Inc.

Note that these results are provided for a manager's development. They don't directly influence the manager's performance ratings or compensation. I actually lost this argument with my team. When we rolled out the Upward Feedback Survey, I thought this would be our opportunity to purge Google of our worst managers, the ones who inflict the most pain on their teams and drag us all down. Stacy Sullivan argued that if we did that, people would start gaming the survey, either pressuring their teams to give higher scores or preemptively firing people who seemed unhappy and likely to give low scores. Stacy and others argued that if we wanted people to be open-minded and change their behaviors, we had to make this a compassionate tool, focused on development rather than rewards and punishment.

She was right. Divorcing developmental and evaluative feedback is essential. We later checked Stacy's instincts, and I was relieved to see that people were using the Upward Feedback Survey as intended: Even in cases where managers gave employees low performance ratings, the employees didn't retaliate by dinging the manager in the next Upward Feedback cycle.

And if managers need help getting better in a specific area, and the checklist isn't doing the trick, they can sign up for the courses we've developed over time for each of the attributes. Taking "Manager as Coach" improves coaching scores an average of 13 percent. Taking "Career Conversations" improves career development ratings by 10 percent, in part by teaching the manager to have a different kind of career conversation with employees. It's not about having the Googler ask for something and the manager promise to deliver. Instead of a transactional exchange, it's a problem-solving exercise that ends with shared responsibility. There's work for both the manager and the Googler to do.

Today, most of our managers share their results with their teams. We don't require it, but we do periodically insert questions into our surveys asking employees if managers have done so. Between our norm of transparency and those small nudges, most managers choose to share. They distribute their reports and then lead a discussion about how to improve their performance, getting advice from their teams. It's a beautiful inversion of the typical manager-employee relationship. The best way to improve is by talking to those providing feedback and asking them exactly what they hope you would do differently.

The first time I shared my results, which were lower than my team's average, I was terrified. I'm in charge of all this stuff for the company! I'm supposed to be the expert! And yet my scores showed that my conduct didn't match my aspirations. My team was 77 percent favorable on me across the fifteen questions we asked that year, which sounds okay until you realize the top quartile managers' teams were 92 percent favorable, and the bottom quartile were 72 percent favorable. For example, I scored particularly low on "My manager helped me understand how my performance is evaluated" at 50 percent favorable, and only 80 percent of my direct reports would have recommended me as a manager to other Googlers.

I made a commitment to my team that I would do a better job of giving clear feedback, spend more time traveling to meet the extended team, and generally work to be a better leader. They appreciated my efforts. Several reached out to encourage me: "Thanks for sharing and setting the right expectations for others. I was one of the few that wrote comments—happy to discuss when we have our next 1:1!" Over time, the team grew happier and functioned better, and my scores improved. I'm still far from perfect (90 percent favorable overall), but managed to get my rating up to 100 percent favorable in giving quality feedback, and 100 percent of my team would now recommend me as a manager.

For Google, the result has been a steady improvement in manager quality. From 2010 to 2012, the average score of managers at Google has improved to 88 percent favorable from 83 percent. Even our lowest-performing managers have gotten better, improving to 77 percent favorable from 70 percent. In other words, our bottom quartile managers have become almost as good as our average was just two years prior. It's actually become harder to be a bad manager. And since we know that manager quality drives performance, retention, and happiness, it means the company will perform better over time.

You may have read all this, leaned back, and concluded: These guys are fooling themselves. People aren't going to give their manager honest ratings. Even if it's anonymous. Even if it doesn't impact pay or promotions. Despite the training everyone gets, and all the effort that goes into recruiting nice people. Human beings just aren't wired that way, and someone, somewhere is going to game the system. Imagine a manager cuts an employee's performance rating. Surely that employee is going to try to get back at the manager by going negative when responding to the UFS?

Well, I have to admit, there is a scrap of truth to this. Mary Kate Stimmler, a member of our PiLab whose PhD studies focused on why people make choices that increase the risk of failure (and who has also won three California State Medals for cracker baking), crunched the numbers and found that, indeed, there is an impact. Under our old 41-point rating scale, a change in employees' performance ratings of plus or minus 0.1 was correlated with them changing their manager's UFS rating by 0.03 on a scale of 0 to 100. In other words, there is an effect, but it's so tiny it doesn't matter at all. Most people, in fact, do the right thing.

Managing your two tails

I've gone into detail on Project Oxygen and the bottom 5 percent for three reasons. First, it's an exceptional illustration of what can be learned and accomplished by focusing on the two tails of performance. Looking at average managers didn't help, nor did benchmarking. Comparing the extremes allowed us to see meaningful differences in behavior and outcomes, which then formed a basis for unceasing improvements in how people experience Google.

Second, it illustrates the notion of compassionate pragmatism. Letting those who are at the bottom of the performance distribution know it, without tying that directly to pay or career outcomes, alerted and motivated them in as positive a way as possible. Hundreds of managers were faced with the fact that they were not good managers. The Upward Feedback Survey replaced gut instinct ("I know I'm a good manager") with real data ("My team is telling me I could be a better manager"). Because of how the results were delivered, and because a great team of HR business partners (who provide not only HR support to hundreds of Googlers, but also act as combination coaches and advocates for individuals), sat quietly with struggling managers when they reviewed their results, most managers responded by asking how they could get better.

Third, and finally, any team can replicate this. I chose to invest precious resources in building the PiLab, at the expense of funding more traditional areas of HR such as training. But there's a shortcut:

1. Care about upgrading your organization. Everyone says they do, but few really take action. As a team leader, a manager, or an executive, you have to be willing to act personally on the results you see, changing your own behavior if needed, and to be consistent over time in staying focused on these issues.

2. Gather the data. Group your managers by performance and employee survey results, and see if there are differences. Then interview them and their teams to find out why. If you're a small team or organization, simply ask people what they value in great managers. Or failing all that, start with our Oxygen checklist.

3. Survey teams twice a year and see how managers are doing. A variety of companies provide survey applications. We rely of course on Google products, specifically Google Sheets, which can send out surveys called Forms and has the advantage of being easy to use, easy to export, and low cost.

4. Have the people who are best at each attribute train everyone else. We ask our Great Manager Award recipients to train others as a condition of winning the award.

Focusing on the two tails is more than anything a result of having constraints: If an organization is investing heavily in staffing, as I hope I've convinced you to do, there are fewer resources left for formal training programs, benefits administrators, and other traditional HR support. Moreover, addressing the two tails is where you'll see the biggest performance improvements: There's little benefit in moving a 40th percentile performer to be a 50th percentile performer, but going from the 5th percentile to the 50th is major.

Studying your strongest people closely and then building programs to measure and reinforce their best attributes for the entire company changes the character of your company. If you also are able to get those who struggle the most to be substantially better, you'll have created a cycle of constant improvement.

Sebastien Marotte, a VP of sales in Europe who had recently joined from Oracle, had a particularly challenging situation:

> My first UFS scores were a disaster; I asked myself, "Am I right for this company? Should I go back to Oracle?" There seemed to be a disconnect, because my manager had rated me favorably in my first performance review, yet my UFS scores were terrible. At Oracle, all that mattered was hitting my numbers. My first reaction was that I had the wrong team. I thought that they didn't understand what we needed to do to win. But then I took a step back and met with my HR business partner. We went through all the comments and came up with a plan. I fixed how I communicated with my team and provided more visibility on our long-term strategy. Within two survey cycles I raised my favorable ratings from 46 percent to 86 percent. It's been tough, but very rewarding. I came here as a senior sales guy, but now I feel like a general manager. [140]

Sebastien today is one of our best and most sought-after leaders.

......

WORK RULES... FOR MANAGING YOUR TWO TAILS

- ☐ Help those in need.
- ☐ Put your best people under a microscope.
- ☐ Use surveys and checklists to find the truth and nudge people to improve.
- ☐ Set a personal example by sharing and acting on your own feedback.

......

9

....

Building a Learning Institution

Your best teachers already work for you....
Let them teach!

American companies spent $156,200,000,000 on learning programs in 2011,[141] a staggering sum. A hundred and thirty-five countries have GDPs below that amount.

Roughly half the money went to programs put on by the companies themselves, and the other half was paid to outside vendors. The average employee received thirty-one hours of training over the year, which works out to more than thirty minutes each week.

Most of that money and time is wasted.

Not because the training is necessarily bad, but because there's no measure of what is actually learned and what behaviors change as a result. Think about it this way. If you spent thirty minutes a week studying karate, you wouldn't be a black belt after a year, but you'd certainly know some basic blocks and strikes. If you spent thirty minutes a week experimenting with pancake recipes, you wouldn't be a Cordon Bleu chef, but you'd be able to make mouthwatering pancakes and be a hero to your friends and family on weekend mornings.[xlix]

[xlix] Seriously. Pancakes are the easiest thing in the world and quick to make from scratch. Mix 2 cups of flour, 1 tablespoon each of baking powder and sugar, and ½ teaspoon of salt. (Actually, double the sugar—yum.) In a new bowl, mix 1 egg and 2 cups of milk. Then mix it all together but stop stirring while there are still air bubbles and small lumps in the batter. Those bubbles will keep your pancakes light and fluffy, and the lumps will cook out. You'll want to stir more. Don't. Trust me.

Yet if you work in the United States, you've on average spent more than thirty minutes a week being trained by your company to do…something. If I look at my years in large and small professional environments, I'd be hard pressed to point to anything I do differently today as a result of training. (The sole exception is the training I received at McKinsey & Company, which conformed to the teaching principles described in this chapter.)

Put another way, the United States spent $638 billion on public education for pre-kindergarten through secondary schools in the 2009–2010 school year,[142] roughly four times what corporations spend on training their people. And yet public schools provide more than ten times as many hours of instruction per learner per year, as well as ancillary developmental programs like sports and academic clubs. And I'd wager that anyone reading this book would agree that they learned more in ten years of school than they did from ten years of corporate training programs.

Why then is so much invested in corporate learning, with so little return?

Because most corporate learning is insufficiently targeted, delivered by the wrong people, and measured incorrectly.

You learn the best when you learn the least

Damon Dunn attended Stanford University in the mid-1990s, before going on to become a professional football player and then founding a real-estate development firm. He recalled heading to a Stanford fraternity party while he was a student.[143] It was 11:00 p.m., dark, and a driving rain pounded the campus. Damon noticed a lone figure at the driving range, methodically hitting golf balls. *Thwack. Thwack. Thwack.*

Mix in some blueberries or banana slices if you wish, and make sure they get coated in batter as well. Cook them in a hot pan with butter. It's just as fast as a premade mix or restaurant, but far, far tastier! This is my own recipe, but it is based on recipes by Mark Bittman and Alton Brown.

Four hours later, at 3:00 a.m., Damon left the party for his dorm. *Thwack. Thwack.* The figure was still there, still hitting golf balls. Damon wandered over.

"Tiger, what are you doing out here hitting balls at three a.m.?"

"It doesn't rain very often in Northern California," replied the kid who went on to become one of the most successful golfers in history. "It's the only chance I have to practice hitting in the rain."

You might expect this kind of diligence from the best athlete in his field. What is fascinating is how narrow the exercise's scope was. He wasn't practicing putting or hitting from a sand bunker. He spent four hours standing in the rain, hitting the same shot from the same spot, pursuing perfection in an intensely specific skill.

It turns out that's the best way to learn. K. Anders Ericsson, a professor of psychology at Florida State University, has studied the acquisition of expert-level skill for decades. The conventional wisdom is that it takes ten thousand hours of effort to become an expert. Ericsson instead found that it's not about how much time you spend learning, but rather how you spend that time. He finds evidence that people who attain mastery of a field, whether they are violinists, surgeons, athletes,[144] or even spelling bee champions,[145] approach learning in a different way from the rest of us. They shard their activities into tiny actions, like hitting the same golf shot in the rain for hours, and repeat them relentlessly. Each time, they observe what happens, make minor—almost imperceptible—adjustments, and improve. Ericsson refers to this as deliberate practice: intentional repetitions of similar, small tasks with immediate feedback, correction, and experimentation.

Simple practice, without feedback and experimentation, is insufficient. I was on my high school's swim team, and among other events competed in the exhausting 200-yard individual medley: fifty yards each of butterfly, backstroke, breaststroke, and freestyle. I joined because I was faster than my friends, and my uncle had played on the Romanian national water polo team, so I guessed

I might have had some natural predisposition. But compared to real swimmers, kids who had been on swim teams since they were six years old and practiced year-round, I was terrible. Between my freshman and senior years, my best times improved by almost 30 percent, but in a typical heat of six swimmers, I'd still have to fight like crazy to come in fifth.

Ericsson could have told me right away what my problem was. I showed up to practice twice a day and swam whatever the coach said to swim, but I couldn't teach myself and was never good enough for the coach to invest even a few minutes in helping me improve my technique. I never experienced deliberate practice. As a result, I got somewhat better, but never had a chance of performing at a high level.

In contrast, McKinsey used to send all second-year consultants to their Engagement Leadership Workshop, a one-week session for about fifty people at a time. The seminars ran throughout the year, rotating between Switzerland, Singapore, and the United States. I, of course, was sent to New Jersey. One of the skills we were taught was how to respond to a furious client. First, the instructors gave us the principles (don't panic, give them time to vent their emotions, etc.), then we role-played the situation, and then we discussed it. Afterward, they gave us a videotape of the role play so we could see exactly what we had done. And we repeated the process again and again. It was a very labor-intensive way of providing the training, but it worked.

Now reflect on the last training program you went to. Perhaps there was a test at the end, or maybe you were asked to work as a team to solve a problem. How much more would you have internalized the content if you'd been given specific feedback and then had to repeat the task three more times?

Building this kind of repetition and focus into training might seem costly, but it's not. As we'll discuss later, most organizations measure training based on the time spent, not on the behaviors changed. It's a better investment to deliver less content and have people retain it, than it is to deliver more hours of "learning" that is quickly forgotten.

The notion of deliberate learning is also relevant to long-term learning. In my town's junior high school, teachers receive tenure after two years of employment. After that, salary increases are purely a function of tenure: There are no meaningful performance standards, and it's almost impossible to be fired. There are no incentives for teachers to compare notes, and they often teach the same subjects for decades. I had an American-history teacher who had been teaching for twenty-five years but hadn't changed the content or his delivery for at least twenty. He had twenty-five years of experience, but twenty of them were the same year repeated twenty times. Repetition without feedback, and in his case also without any motivation. He hadn't improved in two decades.

Unless your job is changing rapidly, this is a universal trap that we all fall into. It's difficult to keep learning and stay motivated when the road stretching ahead of you looks exactly like the road behind you. You can keep your team members' learning from shutting down with a very simple but practical habit. I had the pleasure of working with Frank Wagner, who is now one of our key People Operations leaders at Google, in 1994, when we were consultants. In the minutes before every client meeting, he would take me aside and ask me questions: "What are your goals for this meeting?" "How do you think each client will respond?" "How do you plan to introduce a difficult topic?" We'd conduct the meeting, and on the drive back to our office he would again ask questions that forced me to learn: "How did your approach work out?" "What did you learn?" "What do you want to try differently next time?" I would also ask Frank questions about the interpersonal dynamic in the room and why he pushed on one issue but not another. I shared responsibility with him for ensuring I was improving.

Every meeting ended with immediate feedback and a plan for

what to continue to do or change for next time. I'm no longer a consultant, but I often go through Frank's exercise before and after meetings that my team has with other Googlers. It's an almost magical way to continuously improve your team's performance, and it takes just a few minutes and no preparation. It also trains your people to use themselves as their own experiments, asking questions, trying new approaches, observing what happens, and then trying again.

Build your faculty from within

I can't tell you what to teach your team or organization, since that depends on what your goals are. I can't tell you whether the best way to teach is in person or remotely, through self-study or group classes. That will depend on how your people learn best, and whether they are trying to learn job-specific skills, such as a new programming language, or more general skills, such as how to work better together as a team.

I can, however, tell you exactly where to find the best teachers.

They are sitting right next to you.

I promise you that in your organization there are people who are expert on every facet of what you do, or at least expert enough that they can teach others. We're all familiar with the concepts of maximum and minimum. In theory, you want the best person, the one with the maximum expertise, to be delivering training.

But in mathematics there's a more refined concept: the local maximum. The local maximum is the highest value within a more constrained range of values. The largest number is infinity, but the largest number between one and ten is ten. Yo-Yo Ma is considered the best cellist in the world by many. In South Korea, the very gifted Sung-Won Yang is the most prominent cellist. Yang is the local maximum.

In your company, there is certainly a best salesperson in terms of total sales. By turning to that person to teach others rather than bringing in someone from the outside, you not only have a teacher who is better than your other salespeople, but also someone who understands the specific context of your company and customers. Remember that Groysberg found that exceptional success rarely follows an individual from company to company. Sending your salespeople to the most expensive sales seminars, led by someone who sold products for someone else, is unlikely to revolutionize your sales performance, because the specifics of what your company does matter.

But perhaps you don't want to have your best salesperson teaching. After all, shouldn't she be focused solely on selling? I'd argue that's a shortsighted move, because individual performance scales linearly, while teaching scales geometrically. I'll explain what I mean.

Let's imagine your best salesperson brings in $1 million in sales each year, and that you have ten other salespeople each selling $500,000 per year. Let's further imagine you pull your best salesperson out of the field for 10 percent of her time—five weeks a year—to train the others. She spends those five weeks teaching, following the others around, and generally giving them focused advice as they work to improve small, discrete sales tasks.

Before any training happens, you have revenues of $6 million ($1M + 10 x $500k). In the first year where your best person is training, she brings in only $900,000 because she's teaching instead of selling 10 percent of the time. But if she can improve the other people by just 10 percent, they'll each sell $550,000, and your total firm revenues will now be $6.4 million.

In year two, if there's no further training, your top person sells $1 million but the others are 10 percent better and so sell $550,000 each, for a total of $6.5 million. Your top person spent less time selling for one year, and your revenues went up. Forever. But if instead in year two she took another 10 percent of her time to train people, and they go up another 10 percent to $605,000 each, sales will total $6.05 million. Sales are now up 16 percent for the company—and

21 percent for your new salespeople—in two years. At this rate the new salespeople will double their sales volume in just over eight years (110 percent, 121 percent, 133 percent, 146 percent, 161 percent, 177 percent, and 195 percent at the start of year eight). That rate of increase is geometric growth.[1]

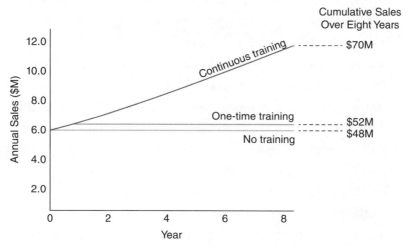

Total company sales under different training scenarios.

And as this rough estimate shows, it's free. The gains among the lower performers more than make up the loss of sales from the top performers.

You don't even need to pull your best salesperson out of the field to make this work. If you break selling down into discrete skills, there may be different people who are best at cold-calling, negotiation, closing deals, or maintaining relationships. The best at each skill should be teaching it.

---

[1] You may notice that the difference in individual sales from year to year keeps increasing. Year two is 10 higher than year one (110 vs. 100), but year eight is 18 higher than year seven (195 vs. 177). Even though the growth rate stays at 10 percent, it's 10 percent of a bigger number each year. If you were to graph revenue growth per year, it would be a curve that bends upward over time at a faster and faster rate.

Former Intel CEO Andy Grove made the same point over thirty years ago:

> Training is, quite simply, one of the highest-leverage activities a manager can perform. Consider for a moment the possibility of your putting on a series of four lectures for members of your department. Let's count on three hours of preparation for each hour of course time—twelve hours of work in total. Say that you have ten students in your class. Next year they will work a total of about twenty thousand hours for your organization. If your training results in a 1 percent improvement in your subordinates' performance, your company will gain the equivalent of two hundred hours of work as the result of the expenditure of your twelve hours.[146]

For the learner, having actual practitioners teaching is far more effective than listening to academics, professional trainers, or consultants. Academics and professional trainers tend to have theoretical knowledge. They know how things ought to work, but haven't lived them. Consultants tend to have shallow and thirdhand knowledge, often gleaned from a benchmarking report from yet another consultant or from spending a few months with one client or another, rather than from sustained experience.

To be fair, it can be valuable to selectively partner with experts, to learn from them and jointly adapt their insights for your company. For example, Marty Linsky of Cambridge Leadership Associates has helped us build our Adoptive Leadership curriculum, and Daniel Goleman helped develop our mindfulness programs, which I'll touch on in a moment. Too often, though, training is outsourced wholesale to outside companies.

It is generally far better to learn from people who are doing the work today, who can answer deeper questions and draw on current, real-life examples. They understand your context better, they are always available to provide immediate feedback, and they are mostly free.

Googler Chade-Meng Tan, or "Meng" as he's popularly known ("I started using the nickname 'Meng' when I realized Americans can't do names with more than one syllable"),[147] was employee number 107, and worked as a software engineer on search from 2000 to 2008 before refocusing his work and life (while still at Google) on achieving world peace through sharing and spreading the notion of mindfulness. Jon Kabat-Zinn, professor emeritus of the University of Massachusetts Medical School, defines it as "paying attention in a particular way; on purpose, in the present moment, and nonjudgmentally."[148] A simple exercise to instill mindfulness is to sit quietly and focus on your breathing for two minutes. It's also been shown to improve cognitive functioning and decision-making.

As an experiment, in late 2013 I invited Googler Bill Duane, a former engineering leader turned Mindfulness Guru, to start my weekly staff meetings with the exercise. I wanted to experiment on ourselves first, and if it worked out we could then try it on larger groups of Googlers, and perhaps eventually the whole company.

The first week was just listening to our breathing, the next was observing the thoughts that ran through our heads as we breathed, working up to paying attention to our current emotions and how they felt in our bodies. After a month, I asked my team if we should continue. They insisted that we did. They told me our meetings seemed more focused, more thoughtful, and less acrimonious. And even though we were spending time on meditation, we were more efficient and were finishing our agenda early each week.

To spread mindfulness principles at Google, Meng created a course, "Search Inside Yourself." Googlers were more receptive to Meng's teachings because he worked for years as an engineer in the

United States and Singapore, so could credibly address the stresses that existed at Google and how mindfulness had changed his life. Meng went on to write a book on the topic and found the Search Inside Yourself Leadership Institute, which he manages while working part-time at Google ("only...40 hours a week," he jokes).[149] His course, book, and institute all work to "develop effective, innovative leaders using science-backed mindfulness and emotional intelligence training."

Bill Duane, who had been a site reliability engineer (translation: He makes sure that Google.com actually works), manages Google's mindfulness team. Bill describes business as a "machine made out of people" and mindfulness as "WD-40 for the company, lubricating the rough spots among driven Googlers."[150] Bill, too, speaks with a gentle authority among other engineers because he has lived what they experience every day.

Meng and Bill aren't alone in deciding that the most important thing they could do—despite their outstanding work as specialists—was to teach. We have a broader program, called G2G or Googler2Googler, where Googlers enlist en masse to teach one another. In 2013, 2,200 different classes were delivered to more than 21,000 Googlers by a G2G faculty of almost three thousand people. Some classes are offered more than once, and some have more than one instructor. Most Googlers took more than one class, resulting in total attendance of over 110,000.

Though teaching takes the G2G faculty away from their day jobs, many courses are just a few hours long and offered only quarterly, so the time commitment for faculty and students is modest. The classes offer a refreshing change of mental scenery, making people more productive when they return to work. And like 20 percent time, G2G makes for a more creative, fun, generative work environment, where people feel deeply invested in what the company does and is. It's a small investment of company resources, with huge dividends.

The content ranges from the highly technical (search algorithm design; a seven-week mini-MBA) to the simply entertaining (tightrope walking; fire breathing; history of the bicycle). Some of the most popular:

- **MindBody Awareness:** Amy Colvin, one of our massage therapists, teaches a thirty-minute class that moves through a dozen qigong (a Chinese practice related to tai chi) poses followed by seated meditation. The course is now available in sixteen cities around the world, often delivered by Hangout. An engineer told Amy, "Being aware of what my physical body needs while my brain is busy coding has helped me significantly reduce stress, not get so worn out, and enjoy my job."

- **Presenting with Charisma:** Sales leader Adam Green teaches Googlers how to move beyond the basics of developing good presentation content to more sophisticated issues like tone, body language, and the use of displacement tactics. An example? "If you find yourself always fiddling with your hands or keeping them in your pockets, try standing behind a chair or podium and planting your hands on the podium so you appear confident. Planting your hands displaces the nervous energy." Another tip: "To rid yourself of saying 'umm' during a presentation, use physical displacement. Every time you are transitioning, do something small but physical, like moving your pen. Making a conscious effort to move your pen will turn your brain off from using a verbal filler instead."

- **I2P (Intro to Programming for Non-Engineers):** Albert Hwang is a leader in our People Operations tools group. He joined Google in 2008 with a degree in economics and taught himself how to program. "I was assigned a project that involved matching up hundreds of Googlers' names with their corresponding office locations and job titles. I quickly realized that a few simple programming scripts could speed up my work and reduce errors. I began to teach myself the programming language Python. My teammates, seeing that my newfound technical skills have saved all of us time, asked me to teach them how to code; thus, in front of a whiteboard in a small conference room, I2P was born." More than two hundred Googlers have since taken Albert's class. In fact, one of his alumni used what they learned to help Googlers find appointments for free on-site flu shots, which helped thousands more Googlers get shots. And because for every on-site vaccination we donate another matching vaccination for meningitis or pneumonia to children in the developing world, the new tool also led to thousands more children receiving vaccinations.[151]

You don't need to create something as formal or widespread as G2G to tap into your teachers. Googlers are given hundreds of other opportunities to learn and teach, any of which could easily be duplicated by motivated employees in your own environment. There are over thirty Tech Advisors, for example, experienced leaders who offer confidential, one-to-one sessions to support Googlers in our technical organizations. These volunteers are selected for their breadth of experience and understanding of Google, and are tasked principally with listening. One of them, Chee Chew, described the experience of being a Tech Advisor this way:

> It's nerve-racking every time. I always feel a lot of anxiety at the top of each session. I have no idea what they're going to ask. There's this whole world of possibility. What if I have nothing to say? . . . As it goes along, very often you feel this sort of connection happen just by listening to what they have to say. I don't have the context and I don't have a strong opinion about what they should do. I don't have a vested interest in the decision, so I listen more and connect more. This is very different from most conversations with my

direct reports and teammates. It's really built for reflection. The connection is with the person as opposed to just the project.

And yet, having a safe, objective person to turn to is sometimes just what people need. Chee continues:

> I remember someone I advised…a high-level woman engineer. She was looking to leave, thought she was at a dead end. Someone convinced her to have a chat with a Tech Advisor. We were signed up to talk for fifty minutes. We talked for two and a half hours. She was able to work through a bunch of stuff. I didn't give a ton of advice. I really listened to her and brainstormed with her, bounced things off her. She came to her own solutions and solved her own problems. She didn't need someone to tell her what to do, only to help encourage her and listen. She's still here.

What surprised me is that in addition to benefiting the person being advised, the advisors themselves benefit as well. Through repeated experience, our company's leaders are building their listening and empathy skills and their own self-awareness. It sounds simple, but the benefits they experience in these sessions have a cascading influence. They claim to be better managers, leaders, and even spouses as a result of building these skills. Note that this is not an HR program, though we manage the administration of it. As Shannon Mahon, the program manager, points out: "The secret sauce is that engineers really own this, not People Operations." Googlers created this for one another.

Similarly, there are volunteers who serve as Gurus, focusing less on individual issues and more on issues of leadership and management across the company. Becky Cotton, at the time part of Google's online payments team, was our first official Career

Guru, a person to whom anyone could turn for career advice. There was no selection process or training. She just decided to do it. She started by announcing in an email that she would hold office hours for anyone who wanted career advice. Over time, demand grew and others volunteered to join Becky as Career Gurus, and in 2013 more than a thousand Googlers had sessions with one of them.

Today we have Leadership Gurus (drawn in part from the annual winners of our Great Manager Awards); Sales Gurus (for advice on selling, so that, for example, a Googler working in the auto industry in Italy can get advice from one in Japan); and Expectant and New Parent Gurus; Having Googlers coach one another not only saves money (I've been told that some external coaches charge $300 per hour or more), but also creates a much more intimate community. As Becky puts it, "You can automate a lot of things, but you can't automate relationships." Becky still coaches 150 people each year, and reports that people often stop her in the halls to tell her, "I would not still be at Google if not for the Guru I spoke to."

Getting started is as easy as *Peanuts'* Lucy van Pelt hanging out her THE DOCTOR IS IN sign. Over the years, Becky has partnered with a number of Fortune 500 technology companies to help them launch their own Guru programs. HR professional Sam Haider and product manager Karen McDaniel, both with financial software company Intuit, did just that. Sam recalls: "We learned of Google's Career Guru program at a Career Development Summit that they hosted and thought it might be a simple, scalable answer to the challenge [of offering 1:1 career advice globally]. We experimented with small groups to validate the idea and then leveraged a grassroots effort already existing in our finance organization. Over the next few months, it grew in popularity and went global."

If you want to unlock your organization's tremendous potential for teaching and learning, you need to create the right conditions. Organizations always seem to have more demand for people development than they can satisfy, and Google is no different. At a global meeting of our development team, one of our sales trainers asked if they would be getting more resources. I told her:

> You won't. Demand for what you can do will always outstrip what you can deliver, because you're doing something that helps people learn and makes them better. You'll always want to do more, since you're a thoughtful, conscientious person. So you'll always be a bit frustrated that you can't do more. Worse, Googlers will always want more from you. Even worse than that, as we grow you'll need to stop doing things that you and your Googlers love, because there will be other, more important things you need to do. You are a precious resource. Our challenge is to figure out together how to help our Googlers teach themselves.

Only invest in courses that change behavior

It's easy to measure how training funds and time are spent, but far more rare and difficult to measure the effect of training. For the past forty-plus years, HR professionals have measured how time is spent, asserting that 70 percent of learning should happen through on-the-job experiences, 20 percent through coaching and mentoring, and 10 percent through classroom instruction.[152] Companies as diverse as Gap,[153] the consultancy PwC,[154] and Dell[155] describe their 70/20/10 development programs on their corporate websites.[li]

But the 70/20/10 rule used by most learning professionals doesn't work.

[li] Note that Google coincidentally used the same proportions as a framework to prioritize our investments from 2005 to 2011. Seventy percent of engineers and resources were deployed on core products such as Search and Ads, 20 percent went to products just outside our core such as News or Maps, and 10 percent went to unrelated programs such as self-driving cars. Sergey Brin developed this approach and Eric Schmidt and Jonathan Rosenberg managed it. Eric, Jonathan, and Alan Eagle detail Google's 70/20/10 approach in their book *How Google Works*.

First, it doesn't tell you what to do. Does the 70 percent mean that people should just figure things out on a day-to-day basis? Or does it refer to rotating people into new jobs so that they learn new skills? Maybe it's about giving people hard projects? Are any of these approaches better than any other one?

Second, even if you know what you're supposed to do, how do you measure it? I've never seen a company that asks managers to log time spent mentoring their teams. Companies can tell how much time and money is spent on classroom training, but beyond that it's often just guesswork. At worst, saying 70 percent of learning happens on the job is a cop-out. It's convenient hand waving that allows HR departments to say that people are learning without proving that they are.

Third, there's no rigorous evidence that allocating learning resources or experiences in this way even works. Scott DeRue and Christopher Myers of the University of Michigan did a thorough review of the literature: "First and foremost, there is actually no empirical evidence supporting this assumption, yet scholars and practitioners frequently quote it as if it is fact." [156]

Fortunately, there's a much better approach to measuring the results of learning programs, and like many great people-management ideas, it's not a new one. In 1959, Donald Kirkpatrick, who was a professor at the University of Wisconsin and past president of the American Society for Training and Development, came up with a model that prescribed four levels of measurement in learning programs: reaction, learning, behavior, and results.

Kirkpatrick's model shares a property of many brilliant ideas: Once it's explained to you, it seems obvious.

Level one—reaction—asks the student for her reaction to the training. It feels great to teach a course and get positive feedback from the students at the end of it. If you're a consultant or professor, people who have a good time and report feeling like they learned are terrific advocates for your class, ensuring future enrollment and revenue. Frank Flynn, a professor at the Stanford Graduate School of Business, once told me the secret to high student evaluations: "Tell lots of jokes and lots of stories. Grad students love stories." He went on to explain that it's a constant trade-off between being engaging and imparting knowledge. Stories key into a human hunger for narrative, rooted in wisdom that's passed from generation to generation through myths and folklore. They are an essential part of effective teaching. But how students feel about your class tells you nothing about whether they have learned anything.

Moreover, the students themselves are often unqualified to provide feedback on the quality of the course. During the class, they should be focused on learning, not on assessing whether the balance of presentation to team exercise to individual exercise is correct.

Level two—learning—assesses the change in the student's knowledge or attitude, typically through a test or survey at the end of the program. Anyone who has taken a driver's license exam has experienced this. This is already a huge improvement on level one, since now we're looking at the effect of the class in an objective way. The drawback is that it's hard to retain newly acquired lessons over time. Worse, if the environment you are returning to is unchanged, the new knowledge will be extinguished. Imagine you just finished a ceramics class, culminating in successfully throwing and firing a glazed pot. If you don't have an opportunity to do it again, you'll lose the new skills you acquired—and you certainly won't refine them.

Kirkpatrick's third level of assessment—behavior—is where his framework becomes powerful. He asks to what extent participants changed their behavior as a result of the training. A few very clever notions are embedded in this simple concept. Assessing behavioral change requires waiting for some time after the learning experience, ensuring lessons have been integrated into long-term memory, rather than hastily memorized for tomorrow's exam and then forgotten. It also relies on sustained external validation. The ideal way of assessing behavioral change is not just to ask the student, but to ask the team around them. Seeking external perspectives both provides a more comprehensive view of the student's behavior and subtly encourages him to assess his own performance more objectively. For instance, if you ask most salespeople how good they are, they will tell you they are among the best in the business. If you ask their clients, and tell them you will ask their clients, you'll get a more modest, and honest, response.

Finally, level four looks at the actual results of the training program. Do you sell more? Are you a better leader? Is the code you write more elegant?

The American College of Surgeons, which uses Kirkpatrick's model, looks for "any improvement in the health and well-being of patients/clients as a direct result of an educational program." [157] Imagine an ophthalmologist who has specialized in LASIK surgery, where a laser is used to reshape the cornea and correct poor vision. It's possible to measure the change in patient outcomes after a physician learns a new technique by recording recovery times, incidence of complications, and degree of vision improvement.

It's much more difficult to measure the impact of training on less structured jobs or more general skills. You can develop fantastically sophisticated statistical models to draw connections between training and outcomes, and at Google we often do. In fact, we often have to, just because our engineers won't believe us otherwise!

But for most organizations, there's a shortcut. Skip the graduate-school math and just compare how identical groups perform after only one has received training.

Start by deciding what your training is supposed to achieve. Let's imagine it's higher sales. Divide your team or organization into two groups, making them as similar as possible. That's hard to do outside of a lab, but at least eliminate obvious differences by having comparable geography, product mix, gender mix, years of experience, and so on. One group then becomes your control group, meaning that nothing changes for them. No classes, no training, no special attention. The second group is your experimental group, and they go through your training.

Then you wait.

If the two groups are truly comparable and the only difference between them has been the training, then any difference in sales results is because of the training.[lii]

What can be counterintuitive and frustrating about this experimental approach is that if you have a problem, you want to fix it for everyone, now. As I shared in chapter 8, managers who take our "Manager as Coach" class improve their coaching scores by 13 percent. We spent a year waiting to see if the class really had an effect, and in the meantime thousands of Googlers were not benefiting from a program that could have helped them.

When I was with another company, each year there would be a new wave of mandatory sales training, guaranteed—we were told—to boost sales performance. But putting everyone through a solution that you think will work doesn't mean it will. A thoughtfully designed experiment, and the patience to wait for and measure the results, will reveal reality to you. Your training program may work, or it may not. The only way to know for sure is to try it on one group and compare it to another group.

Your homework

From the moment we are born, human beings are designed to learn.

But we rarely think about how to learn most effectively.

As a pragmatic matter, you can accelerate the rate of learning in your organization or team by breaking skills down into smaller components and providing prompt, specific feedback. Too many

---

[lii] As long as you're working with comparable groups, another approach is to assign a different type of training to each group. Since you don't have a control group, you can try more things at once. The downside is that you'll have less understanding of what outside factors might be skewing your results. For example, if both groups improve the same amount, is it because both training programs worked or because the economy improved and made selling easier? There's also a chance that differences are due to random variation. There are statistical measures to test this, but a nonquantitative alternative is to run the "experiment" more than once or on additional groups.

organizations try to teach skills that are too broad, too quickly. And measuring the results of training, rather than how much people liked it, will tell you very clearly (over time!) if what you're doing is working.

But we don't just want to learn. We also delight in teaching. You have to look no further than your own family. Every parent teaches, and every child learns. And if you're a parent, you appreciate that often the child is the teacher, and you are the one who learns.

Frank Oppenheimer, the younger brother of noted physicist J. Robert Oppenheimer, purportedly said that "the best way to learn is to teach." [158] He was right. Because to teach well, you really have to think about your content. You need mastery of your subject and an elegant way to convey it to someone else.

But there's a deeper reason to have employee-teachers. Giving employees the opportunity to teach gives them purpose. Even if they don't find meaning in their regular jobs, passing on knowledge is both inspiring and inspirational.

A learning organization starts with a recognition that all of us want to grow and to help others grow. Yet in many organizations, employees are taught and professionals do the teaching.

Why not let people do both?

··

WORK RULES...FOR BUILDING A LEARNING INSTITUTION

- ☐ Engage in deliberate practice: Break lessons down into small, digestible pieces with clear feedback and do them again and again.
- ☐ Have your best people teach.
- ☐ Invest only in courses that you can prove change people's behavior.

··

10

....

Pay Unfairly

Why it's okay to pay two people in the same job
completely different amounts

I never had the pleasure of working with Wayne Rosing, our
first VP of engineering. He retired before my first day. But stories about him still percolate throughout Google. My favorite is about a speech he gave to our engineers in the weeks leading
up to our initial public offering. It was about staying true to our
values, focusing on the user, and how an IPO is just another day.
The next day, we're going to come back to work and keep building
cool stuff for users. People will be richer, some unfathomably so.
But that ought not change who we are. To underscore his point, he
concluded: "If after we go public I see any lamborghinis in our
parking lot, you better buy two of them because I'm going to take a
baseball bat to the windshield of any parked here."

Even though our IPO created many millionaires, we for many
years stayed relatively free of the affectations of conspicuous consumption. This ostentation aversion is as much a reflection of the
historic engineering culture of Silicon Valley as anything special
about Google. *New York Times* journalist David Streitfeld traces
it back to the "founding" of Silicon Valley in 1957,[159] when Robert Noyce, Gordon Moore, Eugene Kleiner, and five others started
Fairchild Semiconductor and developed a way to mass-produce silicon transistors.[160] Streitfeld describes it as a "new kind of company...

one that was all about openness and risk. The rigid hierarchy of the East was eliminated. So was the conspicuous consumption." "The money doesn't seem real," Noyce would later tell his father. "It's just a way of keeping score."[161] The ethos in the Valley has long been "Work hard, but don't show off."

That's of course changed somewhat in recent years, even at Google. The spate of billion-dollar IPOs from companies like Facebook, LinkedIn, and Twitter, combined with the emergence of secondary markets that allow pre-IPO employees to sell shares at multibillion-dollar valuations, has flooded the Valley with money and more than its share of $100,000 Tesla Roadsters and million-dollar homes. Even so, journalist Nick Bilton summarized the current ethos:

> In New York, you see people dressed to impress. In San Francisco, people take pride in wearing a hoodie and jeans to five-star restaurants (despite glossy magazine reports to the contrary).
>
> And in New York people are ostentatious and flashy with their money.
>
> In San Francisco? Sure, Lawrence J. Ellison, Oracle's chief executive and America's Cup champion, is more than happy to show off his wealth. But most wealthy people here tend to hide it, afraid that it won't jibe with the we're-here-to-make-the-world-a-better-place image of Silicon Valley. (I know of one successful founder who owns an old beat-up 1985 Honda that he drives to his secret private jet.)[162]

But Wayne's comments had deeper roots than just giving advice about how to avoid the spasm of gluttony that often accompanies financial success. We had a culture of avoiding ostentation, from the desks we made out of sawhorses and wooden doors to the discarded ski gondolas and monorail cars we reclaimed and used for conference rooms in our Zurich and Sydney offices.[163]

A decommissioned monorail in our Sydney, Australia, office. © Google, Inc.

A decommissioned gondola in the Zurich office. © Google, Inc.

For our products, the purest manifestation of this ethos was a clean, uncluttered search page. It was revolutionary at the time. The prevailing belief was that users wanted a single portal (remember Web portals?) to all of the Internet, with dozens of other portals nested within. Larry and Sergey had a different idea. What if all you had to do was type what you wanted and it magically appeared before you? Here's how our home page compared to those of two leading competitors on February 29, 2000.[164]

Lycos.com home page from circa 2000.

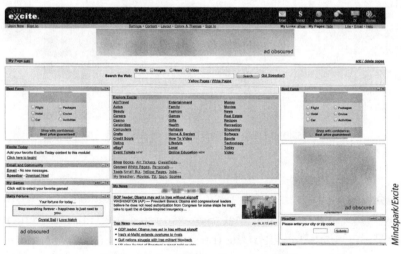

Excite.com home page from circa 2000.

Enter your search terms...

| Google Search | I'm Feeling Lucky |

...or browse web pages by category.

© Google Inc.

Google.com home page from circa 2000. © Google, Inc.

Our spare presentation was so unorthodox that one of our earliest challenges was that users would look at the Google Web page and not type anything. We couldn't figure out why until we went out and did a user study at a nearby college, actually watching students try to use Google. According to Marissa Mayer, at the time a Googler and now CEO of Yahoo, they were so accustomed to cluttered websites that "flashed, revolved, and asked you to punch the monkey" that they thought there had to be more coming.[165] They weren't searching because they were waiting for the page to finish loading. Engineering vice president Jen Fitzpatrick added: "We wound up sticking a copyright tag at the bottom of the page, not so much because we needed a copyright on the page, but because it was a way to say 'This is the end.'" The copyright notice fixed the problem.

Sergey once joked that the reason the Google home page was so empty was that he wasn't very good at HTML, the programming language used to put up Web pages. According to Jen, the reality was that "It became a point of pride and a point of design intent to not throw lots of distractions at you. Our job was to get you from here to there as laser-fast as we could." It was a better user experience: fewer distractions, faster load times, and a quicker path to the destination.[166]

Wayne's careful consideration of how the IPO might change the culture was telling, though, because how to pay people—and how to do so fairly and in consonance with our values—has always been a serious question at Google. In fact, as a management team we have probably spent more time thinking through compensation issues than any other people issue, save recruiting. Recruiting, you'll recall, always comes first, because if you're hiring people who are better than yourself, most other people issues tend to sort themselves out.

For the first year or two, money was tight. But even once we figured out that you could auction advertising on the Internet (imagine!) and revenue started flowing in, we were reluctant to pay high salaries for most of our history. Before Google's initial public offering, our average executive salary was about $140,000. On the one hand, $140,000 is a lot of money. On the other hand, that was for our top people. And the region our employees lived in, Santa Clara and San Mateo counties, had one of the highest costs of living in the country. As a company, our average wage was below the $87,000 median household income in the region.[167]

Almost every new hire took a cut in salary when joining. As I mentioned in chapter 3, we even used this as a recruiting screen, reasoning that only risk-seeking, entrepreneurial types would be willing to take a pay cut of $20,000, $50,000, or even $100,000. New hires were given a further test: They could forgo $5,000 of salary in exchange for 5,000 additional stock options. (Anyone who took that deal would today have an extra $5 million in their pocket.[liii])

As Google grew, we recognized that we'd have to change how we paid people. Low salaries and the promise of IPO-like stock awards wouldn't attract the brightest talent forever. Journalist Alan Deutschman interviewed Sergey on this topic in 2005:

[liii] Our stock split 2-for-1 in 2014, so that early Googler would now hold 10,000 shares of stock.

When there are only a few hundred people in a company, stock is a strong motivation, [Brin] says, because everyone gets enough options to have the chance to really make a lot of money. But "at thousands, it doesn't work that well as an incentive," because there are so many people that the options have to be spread too thin. "And people want the chance to be really well rewarded." Even though Google now has some 3,000 employees worldwide, he says, "I feel the compensation should be more like a startup's [with lower salaries and more stock options]. Not entirely, because there's significantly less risk. But more like one. We provide the upside—maybe not the identical upside, maybe a little less—and higher odds of being successful."

We also wanted to make sure employees stayed hungry and ambitious enough to keep striving for big impact. We studied closely the experiences of other tech companies that had minted millionaires. Deutschman observed that "At Microsoft in the '90s, engineers and marketers took to wearing buttons around the office that said 'fuifv.' You can guess what the first two letters stand for. The last three meant 'I'm fully vested.'"

That was the last thing we wanted!

We spent roughly the next decade making sure that, in addition to having all the right environmental factors and intrinsic rewards in place (our mission, a focus on transparency, a strong Googler voice in how the company operated, freedom to explore and fail and learn, physical spaces that facilitated collaboration), we fine-tuned the extrinsic rewards as well. It came down to four principles:

1. Pay unfairly.
2. Celebrate accomplishment, not compensation.
3. Make it easy to spread the love.
4. Reward thoughtful failure.

A caveat: I'll be throwing around some big numbers in this chapter. Some of this is me rounding off to make the math more self-evident and to avoid getting mired in distracting details. But some of it details the opportunities Google has offered Googlers. Our founders have always been generous. They believe in sharing the value the company creates with employees. As a result, it really is possible to earn, or be awarded, tremendous amounts of money at Google.

Most technology companies of our size stop granting meaningful stock awards to all employees. Instead, they focus on giving bigger awards to executives, as the shares left to the rank and file dwindle to nothing. Outside our industry, one firm I know had a practice of making stock grants worth hundreds of thousands or millions of dollars to the senior executives (the top 0.3 percent), grants worth $10,000 to junior executives (the next 1 percent), and nothing to the other 98.7 percent of employees. Instead of rewarding the best people, they just larded more money onto the most senior ones. I remember one executive who told me privately that he was refusing to retire until his pension hit $500,000 per year (though in his defense, he was also brilliant at his job).

At Google, everyone is eligible for stock awards, at every level of the company and in every country. There are differences in the target award you're eligible for, based on your job and the local market, but the biggest determinant of what you actually receive is your performance. We don't have to include everyone, but we do. It's good business, and it's the right thing to do.

I realize that Google is in a privileged position. I remember working for $3.35 an hour, and how liberating it felt when I later found a job paying $4.25 an hour. And when I got a salaried job that paid $34,000 a year, I felt like I'd never have a financial worry again. After my first paycheck I went out to dinner and for the first time felt flush enough to order an appetizer and a drink with my meal—luxury!

At the same time, companies in lower-margin industries have found that paying people well—even when they don't need to—can be smart business. Costco and Wal-Mart's Sam's Club are both warehouse retailers. Wayne Cascio, of the University of Colorado Denver, compared the two in 2006.[168]

	Sam's Club	Costco
Stores	551	338
Employees	110,200	67,600
Average wage	$10–12/hour*	$17/hour

* Wal-Mart doesn't disclose wages at Sam's Club, but Cascio reports this as a likely range.

Sam's Club vs. Costco.

In addition to offering higher wages, Costco paid 92 percent of the premiums for the 82 percent of employees who had health insurance at the time. Furthermore, 91 percent of employees participated in Costco's retirement plan, with an average company contribution of $1,330 per employee. Despite this much-higher cost structure, and buoyed by a more affluent customer base and more big-ticket items, Costco generated $21,805 in operating profit per hourly-paid employee compared to $11,615 at Sam's Club: 55 percent higher wages but 88 percent higher profit. Cascio explained that "In return for its generous wages and benefits, Costco gets one of the most loyal and productive workforces in all of retailing and...the lowest shrinkage (employee theft) figures in the industry....Costco's stable, productive workforce more than offsets its higher cost."

I'm now choosing to share some of the most sensitive details of how Google has tackled this most private issue, not to flaunt Google's success but because we made a bunch of mistakes in how we approached rewarding people. Along the way, we've studied pay and fairness and justice and happiness. We've learned a bit about how to celebrate success without breeding jealousy. We've applied insights from others and proven that what people think will bring them joy may not always be what does. My hope is that our experiences reveal some lessons that will apply in any work environment—just as they do at Costco—and unlock more freedom, celebration, and satisfaction.

Pay unfairly: Your best people are better than you think, and worth more than you pay them

In a misguided attempt to be "fair," most companies design compensation systems that encourage the best performers and those with the most potential to quit. The first and most critical principle requires you to turn your back on received practice—and it might feel uncomfortable at first.

So-called best practices in compensation start with gathering market data for each job and then designing control limits around how much an employee's individual pay can deviate from market pay and from that of other employees. Typically, companies might allow salaries to vary from market by plus or minus 20 percent, with the absolute best people perhaps 30 percent above market. An average performer might get a 2–3 percent salary increase each year, and an exceptional performer might get 5–10 percent, depending on the company. The perverse result is that if you are an exceptional performer, you will get a series of big increases that then get slower and eventually stop entirely as you approach the upper end of the permissible salary range.

So let's imagine you're doing great work and contributing a lot to your company as a top salesperson, a brilliant accountant, or a clever engineer. In your first year you might get a 10 percent raise, but the next year you'll get 7 percent, and then perhaps 5 percent, and soon enough you're either getting the same increases as an average performer or you've been "red circled" (as the HR people call it) and you won't get salary increases at all! And similar limits apply to how bonuses and stock awards are managed in most places. A well-timed promotion can buy you a bit more time, but you'll soon hit the limit at the next job level.

Something in this system is broken. Most companies manage pay like this to control costs and because they think the range of performance in a single job is somewhat narrow. But they are wrong. Robert Frank and Philip Cook predicted in their 1995 book *The Winner-Take-All Society* that more and more jobs would be characterized by growing compensation inequality, as the best people became increasingly discoverable and mobile, and therefore more able to claim a greater share of the value they create for their employers. This is precisely what the Yankees figured out: The best performers not only command the highest compensation, they will also deliver sustained exceptional results.

The problem is that it's very possible for a person's contribution to grow far more quickly than her compensation. For example, a top-tier consulting firm might pay a new MBA $100,000 per year, and charge clients $2,000 per day ($500,000 per year), roughly five times her salary. In the second year, the MBA might earn between $120,000 and $150,000 and be billed at $4,000 per day ($1 million a year), about eight times her salary. Independent of whether or not this consultant is creating $1 million of value for her employer or client, her share of the value she's creating is dropping each year. This is an extreme example, but it's a pattern that holds in most professional service firms. In fact, economist Edward Lazear of Stanford University has argued that people are on average underpaid relative to their contribution early in their careers, and overpaid later in their careers.[169] Internal pay systems don't move quickly enough or offer enough flexibility to pay the best people what they are actually worth.

The rational thing for you to do, as an exceptional performer, is to quit.

At a Fortune 100 large industrial company, the big "C-level" jobs (the three-initial jobs that start with that key word "chief") turn over about once every five to ten years. If you are an exceptional thirty- or forty-year-old, you get a shot once every decade or so at these top jobs within your company. In the meantime, your pay will move in fits and starts as you get a few big raises and then collide with HR policies that limit your pay until you get the next promotion. For those who are learning and growing most quickly and those performing at the highest level, the one way to make sure your pay is in line with the value you create is to leave the monopolistic internal market and enter the free market. That is, look for a new job, negotiate based on what you are truly worth, then leave. And that's exactly what you see in the labor market.

Why would a company design a system that makes the best and highest-potential people quit? Because they have a misconception of what is fair and lack the courage to be honest with their people. Fairness in pay does not mean everyone at the same job level is paid the same or within 20 percent of one another.

Fairness is when pay is commensurate with contribution.[liv] As a result, there ought to be tremendous variance in pay for individuals. Remember Alan Eustace's argument in chapter 3 that a top-notch engineer is worth three hundred average ones. Bill Gates took a more aggressive view, purportedly saying, "A great lathe operator

---

[liv] The way most companies pay people confuses the notions of equality and fairness. Equality matters tremendously when you're talking about personal rights or justice, but paying everyone equally—or close to equally—results in overpaying your worst people and underpaying your best. HR professionals even have a phrase for this. They say that "internal equity" requires them to constrain the pay of top performers: It wouldn't be "equitable" for some people to make way more money than others. They are technically correct that it wouldn't be equitable, but it would be fair.

A more precise articulation for this chapter would therefore be "pay unequally," but I've opted for "pay unfairly" because it makes the point more starkly. And because massively differentiating compensation will initially feel both unequal (which it will be) and unfair (which it won't be) to HR people and managers.

commands several times the wage of an average lathe operator, but a great writer of software code is worth 10,000 times the price of an average software writer." The range of values for software engineers may be broader than for other jobs, but while a great accountant might not be worth a hundred average accountants, he's surely worth more than three or four of them!

But don't take my word for it. In 1979, Frank Schmidt, of the US Office of Personnel Management, wrote a groundbreaking paper titled "Impact of Valid Selection Procedures on Work-Force Productivity." [170] Schmidt believed, as I've argued in chapters 3 and 4, that most hiring processes fail to select truly talented people. He reasoned that if he could prove that hiring better people resulted in real financial returns, then organizations would focus more energy on doing a better job of hiring.

Schmidt studied mid-level computer programmers working for the federal government. He asked how much more value was created by what he called superior programmers, compared to average programmers, defined as the 85th and 50th percentiles of performance. Superior programmers generated about $11,000 more value each year than average programmers, in 1979 dollars.

He then tried to estimate how much more value would be created for the government if it was better at selecting superior programmers when hiring. His midpoint estimate was about $3 million per year. If as a nation we had been better at screening for all programmers in the country, his midpoint estimate was that an incremental $47 million of value would have been created.

He was wrong about just one thing. Top performers generate way more value than he thought. Alan Eustace and Bill Gates are closer to the truth than Schmidt.

Schmidt assumed that performance was normally distributed. It's not.

Professors Ernest O'Boyle and Herman Aguinis, whom we

met in chapter 8, reported in the journal *Personnel Psychology* that human performance actually follows a power law distribution[171]— pop back to the first few pages of chapter 8 for a refresher. The biggest difference between a normal (also known as a Gaussian) and a power law distribution is that, for some phenomena, normal distributions massively underpredict the likelihood of extreme events. For example, most financial models used by banks up until the 2008 economic crisis assumed a normal distribution of stock market returns. O'Boyle and Aguinis explained: "When stock market performance is predicted using the normal curve, a single-day 10% drop in the financial markets should occur once every 500 years.... In reality, it occurs about once every 5 years." Nassim Nicholas Taleb, in his book *The Black Swan*, made exactly this point, explaining that extreme events were much more likely than most banks' models assumed.[172] As a result, swings and downturns happen far more often than predicted when using a normal distribution, but about as often as you would expect using a power law or similar distribution.

Individual performance also follows a power law distribution. In many fields it's easy to point to people whose performance surpasses their peers' by an inhuman amount. Jack Welch as CEO of GE or Steve Jobs as CEO of Apple and Pixar. Walt Disney and his twenty-six Academy Awards, the most ever for an individual.[173] The Belgian novelist Georges Simenon wrote 570 books and stories (many featuring his detective Jules Maigret), selling between 500 and 700 million copies, and Dame Barbara Cartland of the United Kingdom published more than 700 romance stories, selling between 500 million and one billion copies.[174] (I am clearly writing the wrong kind of book.) As of early 2014, Bruce Springsteen had been nominated for a Grammy forty-nine times, Beyoncé forty-six times, and U2 and Dolly Parton forty-five times each, but are eclipsed by conductor Georg Solti (seventy-four) and Quincy Jones (seventy-nine).[175] Bill Russell of the Boston Celtics won eleven

NBA championships in thirteen seasons,[176] Jack Nicklaus had eighteen major championships,[177] and Billie Jean King won thirty-nine Grand Slam titles.[178]

O'Boyle and Aguinis did five studies encompassing a population of 633,263 researchers, entertainers, politicians, and athletes. The table below compares how many people in each group you would expect to be at the 99.7th percentile of performance using a normal statistical distribution, and how many there are in reality.

	Number predicted by the normal distribution	What you see in reality
Researchers who have published 10 or more papers	35	460
Artists with more than 10 Grammy nominations*	5	64
Members of the US House of Representatives who have served more than 13 terms†	13	172

* The same pattern holds for Oscars, Man Booker Prize nominations, Pulitzer Prize nominations, *Rolling Stone* Top 500 Songs, and thirty-six other awards.
† The same pattern holds in US state and Canadian provincial legislatures, the parliaments of Denmark, Estonia, Finland, Ireland, the Netherlands, and the United Kingdom, and in the New Zealand legislature.

The normal distribution and its failure to predict some types of performance.

When rewarding people within our companies, our intuition leads us to make the same mistake that Schmidt made when studying government programmers. We equate the average with the median, assuming that the middle performer is also the average performer. In fact, most performers are below average:

- 66 percent of researchers are below average in the number of published articles.
- 84 percent of Emmy-nominated actors are below average in total number of nominations.

- 68 percent of US representatives are below average in the number of terms they have served.
- 71 percent of NBA players are below average in the number of points scored.

Note that being below average isn't bad. It's just the math. As the data show, exceptional contributors perform at a level so far above that of most, that they are able to pull the average up well past the median.

The only reason your organization or places like GE have normal performance rating distributions is because HR and management force them to look that way. Companies have expected performance distributions, and raters are trained to hew to them. And that forces pay outcomes to follow the same distribution, which is completely out of line with the value actually created by people.

Applying a proper power law distribution, Schmidt's 85th percentile programmer isn't $11,000 better than average: She's $23,000 better than average. And a 99.7th percentile programmer in 1979 would generate an astounding $140,000 more value than average. Adjusted for inflation, this top 0.3 percent performer generates almost half a million dollars more in value.[179] Alan Eustace's estimate is starting to look pretty reasonable.

O'Boyle and Aguinis break it down: "Ten percent of productivity comes from the top percentile and 26% of output derives from the top 5% of workers." In other words, they found that the top 1 percent of workers generated ten times the average output, and the top 5 percent more than four times the average.

Of course, this doesn't apply everywhere. As O'Boyle and Aguinis point out, "Industries and organizations that rely on manual labor, have limited technology, and place strict standards for both minimum and maximum production" are places where you'll see a normal distribution of performance. In those environments there's little opportunity for exceptional achievement. But everywhere else, this distribution holds sway.

How can you tell if you're in this kind of workplace? Alan gave me a simple test. He asks himself, "How many people would I trade for one Jeff Dean or Sanjay Ghemawat?" Jeff and Sanjay, you'll remember, invented one of the technologies that made Google and just about every big data company that exists today possible.

How many people would you trade for your very best performer? If the number is more than five, you're probably underpaying your best person. And if it's more than ten, you're almost certainly underpaying.

At Google, we do have situations where two people doing the same work can have a hundred times difference in their impact, and in their rewards. For example, there have been situations where one person received a stock award of $10,000, and another working in the same area received $1,000,000. This isn't the norm, but the range of rewards at almost any level can easily vary by 300 to 500 percent, and even then there is plenty of room for outliers. In fact, we have many cases where people at more "junior" levels make far more than average performers at more "senior" levels. It's a natural result of having greater impact, and a compensation system that recognizes that impact.

To make these kinds of extreme rewards work, you need two capabilities. One is a very clear understanding of what impact is derived from the role in question (which requires a complementary awareness of how much is due to context: Did the market move in a lucky way? How much of this was a result of a team effort or the brand of the company? Is the achievement a short- or long-term win?). Once you can assess impact, you can look at your available budget and decide what the shape of your reward curve ought to be. If the best performer is generating ten times as much impact as an average performer, they shouldn't necessarily get ten times the reward, but I'd wager they should get at least five times the reward.[180] If you're adopting a system like this, the only way to stay within budget is to give smaller rewards to the poorer performers, or even the average ones. That won't feel good initially, but take comfort in knowing that you've now given your best people a reason to stay with you, and everyone else a reason to aim higher.

The other capability is having managers who understand the reward system well enough that they can explain to the recipient, and to others who might ask if word were to get out, exactly why a reward was so high and what any employee can do to achieve a similar reward.

In other words, the allocation of extreme awards must be just. If you can't explain to employees the basis for such a wide range of awards, and can't give them specific ways to improve their own performance to these superb levels, you will breed a culture of jealousy and resentment.

Maybe that's why most companies don't bother. It's hard work to have pay ranges where someone can make two or even ten times more than someone else. But it's much harder to watch your highest-potential and best people walk out the door. It makes you wonder which companies are really paying unfairly: the ones where the best people make far more than average, or the ones where everyone is paid the same.

Celebrate accomplishment, not compensation

In November 2004, six years after Google was founded but only three months after the IPO, we gave out our first Founders' Awards.[181] As Sergey wrote in our 2004 founders' letter to shareholders:

> We believe strongly in being generous with our greatest contributors. In too many companies, people who do great things are not justly rewarded. Sometimes, this is because profit-sharing is so broad that any one person's reward gets averaged out with the rewards of everyone else. Other times, it's because contributions are simply not recognized. But we intend to be different. That is why we developed the Founders' Award program over the past quarter.

The Founders' Award is designed to give extraordinary rewards for extraordinary team accomplishments. While there's no single yardstick for measuring achievement, a general rule of thumb is that the team accomplished something that created tremendous value for Google. The awards pay out in the form of Google Stock Units (GSUs) that vest over time. Team members receive awards based on their level of involvement and contribution, and the largest awards to individuals can reach several million dollars....

Like a small start-up, Google will provide substantial upside to our employees based on their accomplishments. But unlike a start-up, we provide a platform and an opportunity to make those accomplishments much more likely to occur.

Against the backdrop of our IPO three months before, Googlers were concerned that those joining just a few months too late were creating similar value for users as those who had joined earlier, but were not being rewarded in the same way. Management too felt that would be unfair. In the back of our minds was the question of whether motivation would drop after the IPO. We wanted to reward and inspire each team by sharing with them a portion of the value they created. And surely nothing could be more exciting, more motivating, than the opportunity to earn millions of dollars?

Two teams, one of which made ads more relevant to users and another that had negotiated a critical partnership, received stock awards worth $12 million in November of 2004.[182] The following year we awarded over $45 million to eleven teams.[183]

As crazy as it sounds, the program made Googlers less happy.

We're a technology company, and the greatest value for users is created by our technical Googlers. Most of our nontechnical staff, all of whom do exceptional things, simply don't have the infrastructure at their disposal to touch over 1.5 billion users every single day. As we launched more and more products, the bulk of

Founders' Award recipients were therefore engineers and product managers. So right off the bat, the half of the company that weren't in technical roles viewed Founders' Awards as demoralizing, since they were unlikely to win.

It turned out that many of the technical people didn't view the awards as attainable either, because not every product has the same impact on the world, rolls out as quickly, or is as easily measurable. Improving our Ads systems has immediate impact that's easy to gauge. Is that more valuable or harder to do than improving the resolution of our Maps imagery? What about building collaborative online word processing tools, like the one being used to write this book? Hard to say. Over time, many technical people started viewing Founders' Awards as just a bit out of reach, reserved mainly for a handful of core product teams.

Within those product areas that won Founders' Awards more frequently, there was always a contentious debate about where to draw the line between those who would be recognized and those who wouldn't. Imagine a multiyear project to launch a new product like Chrome, designed to be the most secure and fastest Internet browser in the world. It's obvious that someone who was on the team for the entire time should get an award, but what if you were on the team for only one year? Shouldn't you get something? What if you were there for just six months? What about someone from the security team who provided valuable input along the way on browser security? What about the marketing professionals who made those wonderful Chrome commercials? (If you're a parent, search for "Dear Sophie." If you manage not to cry, you're a stronger person than I am.) With every award, management did their best to figure out who deserved it, and invariably missed a few people. As a result, each award was accompanied by teeth-gnashing among the near-winners who were working in the right areas but just missed the necessarily arbitrary cutoff.

Ah, but surely the winners were happy?

Not so much. Because of the hype surrounding the program, people assumed every winner received $1 million. In truth, awards could be that high, but most were not. The low end of the range was $5,000. Nothing I would turn down, but you can imagine the shock and disappointment of those who thought they would receive $1 million and then received just one half of 1 percent of that amount.

But surely, surely the few who received $1 million must have been ecstatic?

They actually were pretty happy. I mean, that's just breathtaking. Life-changing.

And then some (though not all) of our best, most creative, most insightful technologists, who had built some of the most impactful products in our history, would realize that they were unlikely to win a Founders' Award for the same work twice, and would immediately try to transfer to new product areas.

Without meaning to, we had created an incentive system that made almost everyone in the company less happy, and even the few happy people ended up wanting to stop doing the essential, innovative work that had earned them the award in the first place!

We quietly went from making these awards every year, to every other year, to even less frequently. There's always the chance we'll do another one, but we haven't for some time.

So, do the shortcomings of this program contradict my earlier advice to pay exceptional people exceptionally well? Actually, no. You should absolutely provide exceptional rewards. But you should do it in a way that's just.

The error we made in our Founders' Awards was that we were celebrating money, even though we didn't mean to. We announced that we were going to offer "start-up-like rewards." We told Googlers that the awards could be up to $1 million. We may just as well have put Googlers in one of these:[184]

Image courtesy of Tessa Pompa and Diana Funk

Luckily, compensation is more equitable and just at Google than in this dream scenario.

Compensation systems are based on imperfect information and administered by imperfect people. They will inevitably have some errors and injustice in the margins. The way we ran the program focused too much attention on the money, which then naturally led to questions of whether the process was just, and to unhappiness.

John Thibaut and Laurens Walker, former professors at UNC–Chapel Hill and the University of Virginia, established the idea of procedural justice in their 1975 book *Procedural Justice* (though I'll grant that they did little to advance the art of snappy book titles).[185] Earlier, the literature maintained that if outcomes were just, people were happy. This was termed distributive justice, meaning that the end distribution of goods, awards, recognition, or whatnot was just.

But that wasn't true in reality. It's like saying you care only about how much a salesman sells, not how they do it. I worked with a salesman at a prior company who terrorized his colleagues, lied to his customers, and consistently blew away his sales quota. Accordingly, he received enormous bonuses. But how he did his work should have mattered just as much as what he accomplished.

Thibaut and Walker called this idea procedural justice. From a distributive perspective, the jerk salesman's getting a huge bonus was just. But his colleagues were furious because, from a procedural perspective, he did everything wrong. And worse, the company implicitly endorsed that behavior by rewarding him.

Kathryn Dekas, a PhD member of our PiLab, described the harm that this situation created: "Fairness perceptions are very powerful. They affect how people think about almost everything at work, but especially how valued they think they are, how satisfied they are with their jobs, how much they trust their supervisors, and their commitment to the organization."

It was only when enough of his colleagues banded together and threatened to quit that the salesman was reprimanded and his conduct improved…somewhat.

Our awards program had inadvertently fallen short of both kinds of justice. We didn't get the lists of winners right, and the range of awards seemed disproportionate to some, lacking distributive justice. Furthermore, the process for deciding winners was opaque, with a perception that over half the company was excluded, failing the procedural justice test too. It's no wonder the program didn't work as we'd hoped.

It's essential that extreme reward systems have both distributive and procedural justice. Recognizing this, or more precisely, having learned this lesson the hard way, we revamped our reward programs. We decided that our public, top-down reward programs would truly be open to the entire company. Rather than just asking our technical leaders for nominations, we turned to our heads of sales, finance, public relations, and other nontechnical departments and encouraged them to nominate teams.

We also shifted these programs from providing monetary awards to experiential awards. This was a profound change for the better. People think about experiences and goods differently than about monetary awards. Cash is evaluated on a cognitive level. A cash award is valued by calculating how it compares to your current salary, or to what you could buy with it. Is it as big as a paycheck or smaller? Can it buy a cellphone or will it help me get a new car? And because money is fungible, as often as not it's spent on staples rather than splurging on a pair of Christian Louboutin shoes or a massage, and fades from memory. Non-cash awards, whether they are experiences (a dinner for two) or gifts (a Nexus 7 tablet), trigger an emotional response. Recipients focus on the fact of what they get to experience, rather than calculating values.[186]

We'd seen this in academic research but were nervous about trying it at Google. When we surveyed Googlers about what they wanted, they unambiguously preferred cash over experiential awards by a margin of 15 percent, and revealed that they would find cash 31 percent more meaningful than experiences. Or more precisely, this is what Googlers thought would make them happiest. But as Dan Gilbert explained in his terrific book *Stumbling on Happiness*, we're not very good at predicting what will make us happy, or how happy it will make us.

So we ran the experiment. For a period of time, in our control groups of Googlers, people who were nominated for cash awards continued to receive them. In our experimental groups, nominated winners received trips, team parties, and gifts of the same value as the cash awards they would have received. Instead of making public stock awards, we sent teams to Hawaii. Instead of smaller awards, we provided trips to health resorts, blowout team dinners, or Google TVs for the home.

The result was astounding. Despite telling us they would prefer cash over experiences, the experimental group was happier. Much happier. They thought their awards were 28 percent more fun, 28 percent more memorable, and 15 percent more thoughtful. This was true whether the experience was a team trip to Disneyland (it turns out most adults are still kids on the inside) or individual vouchers to do something on their own.

And they stayed happier for a longer period of time than Googlers who received money. When resurveyed five months later, the cash recipients' levels of happiness with their awards had dropped by about 25 percent. The experimental group was even happier about the award than when they received it. The joy of money is fleeting, but memories last forever.[187]

We still make exceptional cash and stock awards for exceptional people. And our annual bonus and stock grant awards follow more of a power law distribution. But over the past ten years we've learned that how you determine those awards is as important as how much you award. Programs that failed tests of distributive or procedural justice have been replaced or been improved. We've added a strong emphasis on accumulating experiences, rather than just money. We recognize publicly through experiential awards and reward privately through substantial differentiation in bonus and stock. And Googlers are happier as a result.

Make it easy to spread the love

So far, we've talked about rewards that are given by management, but enlisting employees in providing rewards is important too. As discussed in chapter 6, peers have a much better sense of who is really contributing to a project's success than managers do. Remember Sam from chapter 7, who focused on managing up but whose tactics were painfully evident to his peers? It makes sense, therefore, to encourage peers to reward each other. gThanks (pronounced "gee-thanks") is a tool for making it easy for people to recognize great work.

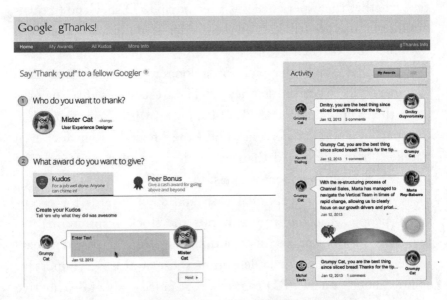

Google's internal recognition tool, gThanks. © Google, Inc.

The simplicity of the design is part of the magic. gThanks makes it easy to send thank-you notes by entering someone's name and then hitting "kudos" and typing up a note. Why is this any better than sending them an email? Because kudos are posted publicly for other people to see and can be shared via Google+. Broadcasting a compliment makes both the giver and the receiver happier. And it's actually fewer keystrokes than writing a private email, which makes it easier to do. To our astonishment, after we launched gThanks we saw a 460 percent increase in the use of kudos compared to a year earlier, when Googlers had to go to a special kudos website, with more than a thousand Googlers visiting the new version every day.

That's not to say the old-fashioned way of doing this is bad. I also keep a "Wall of Happy" outside of my office, where I post the kudos received by members of my team.

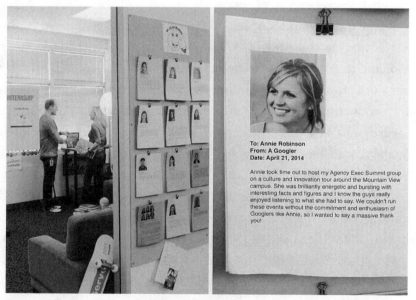

The Wall of Happy outside my office in Google's Mountain View, CA, headquarters.

As Napoleon is purported to have written, though in a more sinister vein: "I have made the most wonderful discovery. I have discovered men will risk their lives, even die, for ribbons!" Simple, public recognition is one of the most effective and most underutilized management tools.

The other feature of gThanks is the peer bonus, which you'll see on the bottom center of the screenshot. Giving employees the freedom to recognize one another is important. Many companies allow employees to nominate an employee of the month, and some allow employees to give modest peer bonuses, with approval from HR or management.

At Google, any employee can give anyone else a $175 cash award, with no management oversight or sign-off required. In many organizations this would be viewed as madness. Wouldn't employees cut side deals to exchange awards? Wouldn't they game the system to earn thousands of dollars in extra income?

That hasn't been our experience.

Over more than a decade, we've only rarely seen abuse of the peer bonus system. And when abuse does happen, it's the Googlers who tend to police it. For example, in the summer of 2013 a Googler sent a note to an internal mailing list looking for volunteers to test a new product. He wrote that anyone who participated would receive a peer bonus as a thank-you. Now, peer bonuses should be granted for above-and-beyond individual contributions, not as payments or incentives. Within an hour, the Googler had sent a second note out to the list. He explained that a colleague had reached out to him and graciously described the intent behind peer bonuses. He confessed that he hadn't been aware that quid pro quos were frowned upon, and he apologized. No harm done.

We've found that trusting people to do the right thing generally results in them doing the right thing. Allowing people to reward one another facilitates a culture of recognition and service, and is a way to show employees that they should be thinking like owners rather than serfs. As Carrie Laureno, a former VP at Goldman Sachs who is now a marketing leader at Google's Creative Lab and the founder of our Veterans Network, explained to me, "Once I got to Google I defaulted to trusting first. More than nine times out of ten, it works out just fine."

And surprisingly, despite the increase in kudos usage, we haven't seen any change in the use of peer bonuses. Making recognition easier has made Google a happier place, and it cost nothing extra.

Reward thoughtful failure

Finally, it's also important to reward failure. While incentives and goals matter, the act of considered risk-taking itself needs to be rewarded, especially in the face of failure. Otherwise, people simply won't take risks.

As David Cote, the CEO of Honeywell, told Adam Bryant of the *New York Times*, "The biggest thing I learned [from working as a commercial fisherman when I was twenty-three] was that hard work doesn't always pay off. If you work on the wrong thing, it doesn't really matter how hard you work, because it's not going to make a difference." [188] Even the best of us fail once in a while. What's important is how you respond to that failure.

Google Wave was announced on May 27, 2009, and launched to the public the following September. It was the output of an exceptional team, working for years to create a product that would leapfrog email, texting, and video chat and create an entirely new way of interacting online.

The tech news site Mashable described it as "Google's biggest product launch in recent memory." [189] Some of the compelling features of Google Wave were:

- It was live. Unlike almost anything available even now, you could see comments and conversations unfold as people typed them, letter by letter. And if you joined a wave late, you could

A look at Google Wave circa 2009 and its innovative interface.
© Google, Inc.

replay the entire conversation as it had happened, giving you the experience of being there.

• It was a platform. Unlike most email or chat products, you could build apps on top of the Wave platform. You could add media feeds, build games, and do just about anything you can do today on most social networks.

• It was open source. The code was open to the public and could be modified and improved.

• It had drag-and-drop functionality. This is now commonplace, but Wave was one of the first social products to allow users to share files and images by simply dragging them onto the screen.

• It had robots. Robots! You could create automated agents that would interact with the conversations in a way that you predetermined. For example, you could design a robot to insert a real-time stock quote any time a company was mentioned.

And yet, the product was a brilliant failure. On August 4, 2010, about a year after launch, we announced we were shutting Wave down. Even though new features were on the way and there was a modestly sized community of users who were passionate about Wave, the rate of adoption had flatlined and our management team decided to pull the plug. Wave was later handed to the Apache Software Foundation,[190] a nonprofit that develops and distributes free open-source software, and some of the team's innovations—like live, concurrent editing—became an integral part of other products.

In addition to pioneering a new product, the Wave team had been run in an experimental way. We were exploring whether setting milestones and allowing teams the possibility of IPO-like rewards for the achievement of IPO-like ambitions would spur greater success. They had chosen to forgo Google bonuses and stock awards for the possibility of much larger rewards. The team had worked for two years on this product, putting in countless hours in an effort to transform how people communicated online. They took a massive, calculated risk. And failed.

So we rewarded them.

In a sense, it was the only reasonable thing to do. We wanted to make sure that taking enormous risks wasn't penalized.

The team of course didn't receive the outsize awards they would have had the product been the runaway success we all wanted. But we made sure they weren't hurt financially by forgoing regular Google compensation. It was less than they'd hoped for, but more than they expected, given the circumstances.

It worked out okay, but not great. The leader of the team resigned, as did several other members. The chasm between what they'd hoped to accomplish and what they did accomplish was too huge. Our financial support helped salve the wound for many, but not for all. Nevertheless, many people stayed and went on to do other marvelous things at Google. The biggest lesson was that rewarding smart failure was vital to support a culture of risk-taking.

Chris Argyris, professor emeritus at Harvard Business School, wrote a lovely article in 1977,[191] in which he looked at the performance of Harvard Business School graduates ten years after graduation. By and large, they got stuck in middle management, when they had all hoped to become CEOs and captains of industry. What happened? Argyris found that when they inevitably hit a roadblock, their ability to learn collapsed:

> What's more, those members of the organization that many assume to be the best at learning are, in fact, not very good at it. I am talking about the well-educated, high-powered, high-commitment professionals who occupy key leadership positions in the modern corporation....*Put simply, because many professionals are almost always successful at what they do, they rarely experience failure. And because they have rarely failed, they have never learned how to learn from failure....*[T]hey become defensive, screen out criticism, and put the "blame" on anyone and everyone but themselves. In short, their ability to learn shuts down precisely at the moment they need it the most.[192] [italics mine]

A year or two after Wave, Jeff Huber was running our Ads engineering team. He had a policy that any notable bug or mistake would be discussed at his team meeting in a "What did we learn?" session. He wanted to make sure that bad news was shared as openly as good news, so that he and his leaders were never blind to what was really happening and to reinforce the importance of learning from mistakes. In one session, a mortified engineer confessed, "Jeff, I screwed up a line of code and it cost us a million dollars in revenue." After leading the team through the postmortem and fixes, Jeff concluded, "Did we get more than a million dollars in learning out of this?" "Yes." "Then get back to work." [193]

And it works in other settings too. A Bay Area public school, the Bullis Charter School in Los Altos, takes this approach to middle school math. If a child misses a question on a math test, they can try the question again for half credit. As their principal, Wanny Hersey, told me, "These are smart kids, but in life they are going to hit walls once in a while. It's vital they master geometry, algebra one, and algebra two, but it's just as important that they respond to failure by trying again instead of giving up." In the 2012–2013 academic year, Bullis was the third-highest-ranked middle school in California. [194]

Your leap of faith: putting the four principles into practice

Some pretty stratospheric numbers made an appearance in this chapter, and I know that in most of the real world these kinds of pay levels are far out of reach.[lv] In fairness, it's not that common

---

[lv] Clear exceptions include Frank and Cook's "winner-take-all markets," where the differences between the very best people and the next best are more readily observable, like professional sports, music, or acting. In those markets you see top pay in the tens of millions. And it follows a power law distribution. The Screen Actors Guild (SAG), for example, hasn't released member compensation statistics since 2008, but it's possible to piece together the distribution from various articles. Very roughly, the bottom third of active SAG members made no money from acting in 2007, and the next third earned less than $1,000. The next group, between the 68th and 95th percentiles, were paid between $1,000 and $100,000. The 94th to 99th percentile actors earned between $100,000 and $250,000. And the top 1 percent earned over $250,000. The top 1 percent of the top 1 percent earned even more: Will Smith was the highest-paid actor, with over $80 million in earnings, followed by Johnny Depp ($72 million), Eddie Murphy and Mike Myers ($55 million each), and Leonardo DiCaprio ($45 million). [Sources: "A Middle-Class Drama," *Los Angeles Times*, May 28, 2008; "SAG Focuses Hollywood Pitch," *New York Times*, July 1, 2008; "Dues for Middle-Class Actors," *Hollywood Reporter*, March 3, 2012; "Hollywood's Best-Paid Actors, *Forbes*, July 22, 2008.]

at Google either, despite competing for great people in one of the most competitive global talent markets.

That said, the underlying notion that performance follows a power law distribution would have been true at almost any place I've worked, whether it was at a public school, a charitable nonprofit, a restaurant, or in consulting. In every environment, there were more exceptional people than you would expect based on normal performance management curves. And the exceptional people were clearly, obviously, far better than everyone else. Teachers who won awards every year, fund-raisers who brought in three times as much as the next person, waiters who would (maddeningly) get twice the tips I did every single night. And they were always paid "fairly," which meant they couldn't earn too much more than the average performers because the average performers might get offended. The truth was, we could all see how much better they were, and how much more they deserved. If your best person is worth ten of the average people, you must pay "unfairly." Otherwise, you're just giving them a reason to quit.

At the same time, when you do reward people, make sure to sprinkle in experiences, not just cash. Few people look back on their lives as a series of paychecks. They remember the conversations, lunches, and events with colleagues and friends. Celebrate success with actions, not dollars.

Trust your people enough to let them recognize each other, as well. It may be kudos and nice words, or it may be small awards. A gift card for a local coffee shop or a bottle of wine sent to an understanding spouse as a thank-you for the employee working late. Give employees the freedom to care for each other.

And if people shoot for the stars and only hit the moon, don't treat them too harshly. Ease the pain of failure to leave room for learning. As Larry often says: If your goals are ambitious and crazy enough, even failure will be a pretty good achievement.

WORK RULES...FOR PAYING UNFAIRLY

☐ Swallow hard and pay unfairly. Have wide variations in pay that reflect the power law distribution of performance.
☐ Celebrate accomplishment, not compensation.
☐ Make it easy to spread the love.
☐ Reward thoughtful failure.

The Best Things in Life Are Free (or Almost Free)

Most of Google's people programs can be duplicated by anyone

People can exist, indeed did exist for thousands of years, without companies. But companies can't exist without people. In tough economic times we lose sight of that fact. Companies struggle to maintain their profit margins or even to keep their doors open, cutting working hours and benefits. People take what jobs they can. Work becomes ever more miserable. Then companies are surprised that attrition leaps as soon as the economy improves.

In contrast, our founders' letter from our 2004 IPO filing read:

> We provide many unusual benefits for our employees, including meals free of charge, doctors and washing machines. We are careful to consider the long-term advantages to the company of these benefits. *Expect us to add benefits rather than pare them down over time.* We believe it is easy to be penny wise and pound foolish with respect to benefits that can save employees considerable time and improve their health and productivity. [italics mine]

As we've added programs, I've been pleased to find that the ones that matter the most to Googlers don't come with enormous price tags. In part this is because the times when a person could most use his employer's help come infrequently, as I'll explain shortly. And in part it's because adding new programs is largely about saying yes to employee ideas.

Most people assume Google spends a fortune on doing special things for our employees.

Aside from our cafés and shuttles, we don't.[lvi] Most of the programs we use to delight and care for Googlers are free, or very close to it. And most would be easy for almost anyone to duplicate. The astonishing thing is that more companies don't come up with ones of their own. All it takes is imagination and the will to do it.

We use our people programs to achieve three goals: efficiency, community, and innovation. Every one of our programs exists to further at least one of these goals, and often more than one.

Encouraging efficiency in your professional and personal lives

Most companies want workers to be efficient. Google is no different. As you might guess, we measure everything: We closely monitor how efficiently our data centers are used, computer code quality, sales performance, travel expenses, and so on. We also want people to be efficient in their personal lives. Googlers work hard, and nothing is more dispiriting than finishing a grueling week of work and

---

[lvi] We provide free meals for all our employees and their guests with a goal of creating community and opportunities for people to generate new ideas. In 2013 we served over 75,000 free meals every day. Our shuttle service ferries Googlers from shuttle stops around the San Francisco peninsula to our offices in San Francisco and Mountain View. In 2013 our Wi-Fi-equipped buses drove 5,312,156 miles, making us the largest private mass transit provider in the area. The shuttles make Googlers' lives more efficient by shortening commuting time and making the time useful. They also take tens of thousands of cars off the road that would otherwise be clogging the highways.

coming home to time-consuming, mundane chores. So we offer on-site services to make life easier. They include:

- ATMs
- bike repair
- car washes and oil changes
- dry cleaning, where Googlers drop clothes off in bins and pick them up a few days later
- farm-fresh and organic produce and meat delivery
- holiday fairs, where merchants come on-site to sell products
- mobile haircuts and beauty salons that pull up in enormous buses outfitted with barber's chairs
- mobile libraries, a service provided by many of the towns where we have offices

These cost Google nothing, because we don't pay for them. Entrepreneurs want to provide these services and require only our permission to come on our site. Googlers pay for the services (though in some cases we are able to negotiate volume discounts on their behalf). And in some cases, such as grocery delivery, Googlers themselves organize the services.

And they are easy to establish. In our Chicago office, a Googler asked a local nail salon owner if she would set up shop in a conference room each week so Googlers could get their mani-pedis at the office. Now it's a Googler-maintained perk that costs Google nothing more than the coffee the manicurist drinks. All we needed to provide was the culture, where Googlers knew they could suggest new programs and shape their own workplaces.

There are some services that do cost Google money, but the amounts are relatively modest and the impact on Googlers is immense. For example, for those who bike or take public transportation to work, we maintain a handful of electric vehicles for their use in case they need to pick up groceries or a friend at the airport.

There is also a concierge team of five people who support our more than fifty thousand employees by helping with travel planning, finding plumbers and handymen, ordering flowers and gifts, and otherwise saving Googlers an hour or two of time as they can. To be very clear, these are costs Google can afford because we're large enough that a few extra people or vehicles (the cost of which is depreciated over years) are not a large percentage of our cost structure. But we also have online bulletin boards that anyone could duplicate, where people offer tips about local services like plumbers, give tutor recommendations, and share local deals they've spotted. And once you have fifty or a hundred employees to make up a potential market, you can start negotiating volume discounts with local businesses.

A community that spans Google and beyond

A sense of community helps people do their best work just as surely as increasing efficiency does by sweeping away minor chores and distractions. As we've grown, we've fought to maintain the sense of community we had when we were just a handful of people, and we've expanded our definition of community to include Googlers' children, spouses, partners, parents, and even grandparents. Many companies hold a "Take Your Child to Work Day," as we have for many years. In 2012, we held our first annual "Take Your Parents to Work Day," and welcomed over two thousand parents to our Mountain View office and over five hundred parents to our office in New York. Each day starts with a welcome, and then either a peek at the future of what we're building or an insider's account of our history. One year we had Omid Kordestani, our founding sales executive, talk about growing Google from ten people to twenty thousand. Another year, Amit Singhal, our SVP of Search, recalled how, as a child in India, he watched *Star Trek*'s Captain Kirk direct his computer by talking to it, and how astonishing it is that Google Now allows him to do exactly the same thing. The rest of the day is filled with product demonstrations, where parents can check out our self-driving cars or stand in a twenty-foot-tall room with Google Earth projected all around them, explore the campus, and then join a special TGIF hosted by Larry and our senior team. We now host these days in more than nineteen offices, including Beijing, Colombia, Haifa, Tokyo, London, and New York City, and add more each year.

Take Your Parents to Work Day isn't about humoring helicopter parents who continue to coddle their fully grown children. Instead, it's a chance for us to say thank you and broaden the Google family. Not surprisingly, our parents are incredibly proud of us and, surprisingly, most of them have no idea what we do for a living. Helping them appreciate the impact their children have, even when those children are fifty years old, is heartwarming. I was stopped a dozen times by parents with tears in their eyes, delighted at the chance to get closer to their children, and grateful to be recognized for having raised such amazing people. The Googlers loved it too. Tom Johnson wrote that "Thinking about taking [my mother] around the place I am so proud to work [at] and seeing how happy she was to spend the time with me brings a wide smile to my face every time."

It was my favorite day at Google ever.

We also work to create a community within the company. As discussed in chapter 2, the Q&A portion of TGIF is the most critical part of the meeting, as any Googler can ask any question, ranging from "Why is my chair so uncomfortable?" to "Are we sufficiently sensitive to user concerns about privacy?" Events like the gTalent shows, where you suddenly realize a saleswoman is also a champion equestrian acrobat (that's doing gymnastics on the back of a moving horse!) and an engineer is a nationally ranked ballroom dancer, or Random Lunches, where people are set up with Googlers they've never met to get to know each other over lunch, are easy to co-ordinate and make the place seem smaller and more intimate. These programs cost almost nothing except the time spent dreaming them up (though we do offer snacks and drinks at some of them—that's optional).

We have more than two thousand email lists,[lvii] groups, and clubs at Google, ranging from unicycling and juggling clubs (which seem to be something that every technology company is required to have) to book clubs, financial planning groups, and even one jokingly called Fight Club after the Brad Pitt movie. They don't really fight. Some wag just thought that if you're going to have clubs, you of course should have a Fight Club. (I can't really talk about it.) Among our clubs, the Employee Resource Groups (ERGs) are worth special mention. Today we have more than twenty, many of which enjoy global membership, including:

- American Indian Network
- Asian Googler Network
- The Black Googler Network
- Filipino Googler Network
- Gayglers (focused on issues facing lesbian, gay, bisexual, or transgender people)
- Greyglers (for older Googlers)
- HOLA (Hispanic Googler Network)
- Indus Googler Network
- The Special Needs Network (examples of special needs might include autism, ADHD, or blindness)
- VetNet (Veterans Network)
- Women@Google

[lvii] Among the questions and topics that come up: a father needing help putting together a sheep costume at the last minute for his daughter; a Googler offering samples of his homemade beef jerky; people looking to borrow wedding cake stands, winter gloves, and—once—a sword; lost and found; referrals for legal advice and preschools; executives looking for rides to the airport; and warnings about mountain lions near the office.

Over a decade ago I noticed that Time Inc., the publishing company, had their own versions of these organizations. One was an Asian American Club, and they'd put up a flyer advertising a class on feng shui they were hosting, and inviting everyone in the company to come. I was struck by this, since my prior experiences had been with groups that focused on serving their own communities but rarely on building connections across communities.

Similarly, at Google all these groups come together in various ways. Anyone can join any ERG. We have numerous Mosaic chapters that bring people together across all the ERGs in offices that are too small to support dedicated VetNet or Greygler groups. There's a series of Sum of Google conferences and events, ranging from potluck dinners and movie nights to career talks and volunteer projects, premised on the idea that many of these groups have common ground in what they experience in society.

Other noteworthy groups include our fifty-two Culture Clubs, which in addition to keeping our culture strong in each office, organize events that both bring Googlers together and forge deeper connections with people outside Google. A few recent examples from programs run by volunteers, ERGs, or Culture Clubs:

- Almost two thousand Googlers joined pride marches in the United States in 2014, and hundreds more participated in Hyderabad, São Paulo, Seoul, Tokyo, Mexico City, Paris, and Hamburg.

- The Hispanic Googler Network held a family health day in Mountain View, welcoming more than three hundred local, low-income families to our campus to share information about technology, fitness, and nutrition. Doctors and nutritionists were on hand to provide medical advice and referrals for ongoing support.

- The Black Googler Network (BGN) hosts an annual out-reach trip. In 2014, thirty-five members from seven offices met in Chicago for three days of outreach to minority-owned small businesses, career development, and community partnership. They were able to reach thirty minority business owners by hosting "small business blitz" office hours, where business owners could pick Googlers' brains at six different stations, covering topics ranging from social marketing to building a website. BGN also hosted more than forty elementary and high school students, including at-risk youth, at the Google Chicago office. The students received a tour, presentations on opportunities and diversity in computer science, and a hands-on experience with Blockly, an introductory coding activity.

- Each month in Singapore, Googlers spend two afternoons supporting abused women and men from all over Asia who have lost their jobs or need help getting back on their feet in some way. Googlers teach the aspiring businessmen and businesswomen how to use the Internet and Google products to gain the skills and confidence required to find a new job or start a business in their home country.

- VetNet helps veterans build skills and find jobs when they return to civilian life. One recent example was "Help a Hero Get Hired," a resume workshop for veterans transitioning out of the military. VetNet ran fifteen workshops in twelve cities across the country as part of our 2013 GoogleServe week of community service.

- In support of a Googler in our Amsterdam office who was waiting for a kidney transplant, the office organized a "Pay to Pee Day," where every time nature called, Googlers were asked to pay a modest fee that was then donated to the Dutch Kidney Foundation.

- Googlers in Tokyo held a "Sell Your Soul" auction in Tokyo to raise money in the wake of the 2011 tsunami. Googlers offered services that reflected their deepest selves, ranging from cooking tips and coding advice to providing a guided 700-kilometer cycling trip to northern Japan. They raised $20,000 for tsunami relief.

- In Mountain View, California, Googlers lead computer and English classes for members of our janitorial staff as part of a program called Through Education And Dialogue.

- Our Madrid office, responding to record unemployment in Spain, decided to donate a ton of food over a forty-day period, which would provide seven thousand hot meals to those in need. The team ultimately collected four tons of food, which was matched by Google and donated to Cáritas, a local aid organization.

And lest you think Google is some kind of forced-march funhouse, of the kind described in Dave Eggers's dystopian novel *The Circle*,[195] there's no expectation or requirement that people need to do any of this stuff. Just like when we were all kids in school, some people join clubs, some play more than others, and some just want their own quiet place to get their work done.

In chapter 9 we talked about how learning works at Google, but as so often happens, we didn't predict an elusive but powerful side benefit of these programs. In 2007 we created our Advanced Leadership Lab, a three-day program for senior leaders where we deliberately assembled a diverse group, spanning a range of geographies, professional functions, genders, social and ethnic backgrounds, and tenures. Stacy Brown-Philpot, at the time a director in our sales organization, who went on to be an entrepreneur-in-residence at Google Ventures before becoming chief operating officer of TaskRabbit, was in the first session. Years later, she and I compared notes on what a special experience it was to build that program from scratch.

"I loved the people I met. I didn't realize that we had so many amazing people doing so many different things," she told me.

"Who do you stay in touch with?" I asked.

"No one."

"But…"

"It's weird. I've never had a need to reach out to them. But I feel better just knowing they're there."

Stacy's comment took me back to my own childhood, and an exchange between Pooh and his dear friend Piglet in A. A. Milne's *Winnie-the-Pooh*:

> Piglet sidled up to Pooh from behind. "Pooh?" he whispered.
> "Yes, Piglet?"
> "Nothing," said Piglet, taking Pooh's hand. "I just wanted to be sure of you."

Perhaps some of the value of these networks and groups comes from simply knowing they are there.

Fueling innovation

These efforts do genuine good in the world, but they also make a difference inside our own walls. When people gather together in unexpected ways, it inevitably spurs innovation—the third goal driving our programs. Amazon lists 54,950 books on innovation for sale, presenting many competing and often conflicting theories. Google, of course, has a number of approaches, but the most salient one is the way we use our benefits and also our environment to increase the number of "moments of serendipity" that spark creativity.

David Radcliffe, our VP of Real Estate and Workplace Services, lays out our cafés and manages the lengths of lines so that there are "casual collisions" between people who might have interesting conversations.

We dot our floors with microkitchens, pockets where you can grab a coffee, a piece of organic fruit, or a snack, and take a few minutes to relax. Often you'll see Googlers chatting and comparing notes over a cookie and a chessboard or around a pool table. Sergey once said, "No one should be more than two hundred feet away from food," but the real purpose of these microkitchens is to do the same thing Howard Schultz tried to create with Starbucks. Schultz saw the need for a "third place" beyond the home and office, where people could relax, refresh, and connect with one another. We try to do the same thing, by giving Googlers a place to meet up that looks and feels different from their desk. And we use the placement of these microkitchens to draw people from different groups together.

Google's microkitchens are interspersed throughout our offices.
This is a particularly nice one. © Google, Inc.

Often they'll sit at the border between two different teams, with the goal of having those people bump into one another. At minimum, they might have a great conversation. And maybe they'll hit on an idea for our users that hasn't been thought of yet.

Ronald Burt, a sociologist at the University of Chicago, has shown that innovation tends to occur in the structural holes between social groups. These could be the gaps between business functional units, teams that tend not to interact, or even the quiet person at the end of the conference table who never says anything. Burt has a delicious way of putting it: "People who stand near the holes in a social structure are at higher risk of having good ideas."[196]

People with tight social networks, like those in a business unit or team, often have similar ideas and ways of looking at problems. Over time, creativity dies. But the handful of people who operate in the overlapping space between groups tend to come up with better ideas. And often, they're not even original. They are an application of an idea from one group to a new group.

Burt explains: "The usual image of creativity is that it's some sort of genetic gift, some heroic act.... But creativity is an import-export game. It's not a creation game.... Tracing the origin of an idea is an interesting academic exercise, but it's largely irrelevant.... The trick is, can you get an idea which is mundane and well known in one place to another place where people would get value out of it."

These gently orchestrated encounters aren't our only trick. We also try to constantly feed new thinking and ideas into the organization. Employees are encouraged to give Tech Talks, where they share their latest work with anyone who is curious. We also bring in star thinkers from outside. Susan Wojcicki and Sheryl Sandberg, a sales VP at the time and now COO of Facebook, were instrumental in growing the concept behind these talks, using their networks and interests to recruit a range of speakers to Google to speak about leadership, women's issues, and politics.

Googlers first self-organized these events into a more formal program in 2006, when they noticed more and more authors visiting to speak with our book-scanning teams. The volunteers asked visiting authors to stick around for a conversation, and our first official Authors@Google guest was none other than Malcolm Gladwell.

This grew into today's broader program called Talks at Google, a speaker series where authors, scientists, business leaders, performers, politicians, and other thought-provoking figures are invited to campus to share their thoughts. Supported by more than eighty volunteers, including terrific people like Ann Farmer and Cliff Redeker, more than two thousand speakers have visited Google. Guests have included Presidents Obama and Clinton, Tina Fey, *Game of Thrones* author George R. R. Martin, Lady Gaga, economist Burton Malkiel, Geena Davis, Toni Morrison, George Soros, microfinance pioneer Muhammad Yunus, Questlove, Anne Rice, Noam Chomsky, David Beckham, and Dr. Oz. More than 1,800 of these talks have been recorded and collectively watched more than 36,000,000 times on YouTube,[lviii] with more than 154,000 followers. Not bad for a bunch of volunteers using 20 percent time. Ann describes the ultimate goal as "taking creative ideas from outside, mingling them with a passionate Google audience, and amplifying the conversations to the billions of YouTube users worldwide. As Malcolm would have said, we want to be connectors."

Add this to the dozens of weekly Tech Talks about internal Google topics, and these visitors and conversations create an atmosphere of constant, bubbling creativity and stimulation, while also giving people a break from their day-to-day work to recharge their imaginations.

A program of this scale might seem out of reach to a small orga-

---

[lviii] You can watch them at http://www.youtube.com/user/AtGoogleTalks.

nization, but there's no big difference between how we started and what anyone can do. Not every company can get Noam Chomsky to visit, but anyone can call a local college and ask a literature professor to come give a talk about David Foster Wallace, ask a string quartet to perform a piece at lunchtime, or have someone demonstrate how the Alexander Technique will cut down on backaches at the desk. And it costs nothing.

There are unforeseeable benefits as well. Meng Tan again:

I hosted a talk on mindfulness meditation by Jon Kabat-Zinn, which has been watched 1.8 million times the last I checked. One viewer emailed me to tell me that video saved his life. He was about to kill himself. He randomly chanced upon that video...and started practicing mindfulness. His depression went away, his addiction was cured, he found a job he loved, had six promotions in six years after that, and is now in a fulfilling relationship. He emailed me six years after watching the video to let me know.

Find ways to say yes

It's reasonable to wonder, with all this stimulation and activity going on, when do people actually work?

It's true that if a single person partook of all these activities, it would consume their entire day. In reality, no one uses all the services or attends every talk, just like no one uses 20 percent time all the time. I've never used our dry cleaning service, but there are people who use it every week. I do get my hair cut in one of our buses: Max and Kwan do a terrific job,[197] and twenty-five minutes later I'm back to work.

In a way, it's like going to a shopping mall. There are many stores you'll never enter. But there's something for everyone:

Program	Cost to Google	Cost to Googler	Benefit to Googlers or Google
ATMs	Free	Free	Efficiency
Bureaucracy Busters	Free	Free	Efficiency
gTalent Show	Free	Free	Community
Holiday fairs	Free	Free	Efficiency
Mobile libraries	Free	Free	Efficiency
Random Lunch	Free	Free	Community; innovation
TGIF	Free	Free	Community
Bike repair	Free	Yes	Efficiency
Car wash and oil change	Free	Yes	Efficiency
Dry cleaning	Free	Yes	Efficiency
Haircuts and salons	Free	Yes	Efficiency
Organic grocery delivery	Free	Yes	Efficiency
Concierge	Negligible	Free	Efficiency
Culture Clubs	Negligible	Free	Community
Employee Resource Groups	Negligible	Free	Right thing to do; community; innovation
Equality in benefits	Negligible	Free	Right thing to do
gCareer (return to work program)	Negligible	Free	Right thing to do; efficiency
Massage chairs	Negligible	Free	Efficiency
Nap pods	Negligible	Free	Efficiency
Onsite laundry machines	Negligible	Free	Efficiency
Take Your Child to Work Day	Negligible	Free	Community
Take Your Parent to Work Day	Negligible	Free	Community
Talks @Google	Negligible	Free	Innovation
Loaner electric vehicles	Modest	Free	Efficiency
Massage	Modest	Yes	Efficiency
Free food	High	Free	Community; innovation
Shuttle service	High	Free	Efficiency
Subsidized child care	High	Yes	Efficiency

A sampling of Google's perks. © Google, Inc.

Almost everything we do is free or low cost. All of these programs work to create efficiency, community, or innovation. Some might argue that this is a gilded cage, a trick designed to convince Googlers to work more or stick around longer. But that fundamentally misunderstands not just our motives, but also how work happens in companies like ours.

The reason I can't point to statistics about the economic value created by having free laundry machines on-site is because I don't really care. I remember the hassle early in my career of stockpiling quarters and lugging boxes of detergent up and down the stairs from my apartment to the shared washing machine in the basement, then being stuck at home for hours lest someone come along and steal my shirts. Super annoying. Why wouldn't we throw a few machines and detergent in a spare room on our campus to make life a little more pleasant? Why wouldn't we have speakers come on campus and give talks? After all, I get to enjoy those presentations as well. And when Dr. Oz visited, Eric (our CEO at the time) and I invited him up to join our management meeting... but only after he'd signed at least three hundred books for his Googler fans.

More significantly, Google isn't some sweetly baited trap designed to trick people into staying at the office working all the time. Why would we care how many hours people work, if their output is good? And the reality is that where you work shouldn't matter at all. It's absolutely necessary to have teams come together, and we get great product ideas and partnerships resulting from people bumping into each other. But do I want people in the office from 9:00 a.m. to 5:00 p.m.? Is there any reason I'd want them there earlier? Or later? People should and do come and go as they please. Many engineers don't roll in until 10:00 a.m. or later. After they head home, there's another burst of activity online in the evenings as people log back in. It's not up to us to tell people when they should be creative.

Are we coddling Googlers so much that they'll never look for another job? Doubtful. We conduct exit surveys, and no one has ever said the perks made them stay, or that they were joining us because of them. There's no big secret here: We do all this because it's (mostly) easy, rewarding, and it feels right.

But is it really practical to say anyone can do this?

Remember that most of these programs are free. They simply require someone at your company to go out and find a vendor who wants to sell to your employees, or organize a lunch, or invite a speaker to visit. Everyone wins.

And some of the programs that once were esoteric are becoming commonplace. Yahoo now has a program called PB&J, short for Process, Bureaucracy, and Jams,[198] which looks a lot like Bureaucracy Busters. Twitter, Facebook, and Yahoo all have versions of TGIF. The idea of all-employee meetings certainly wasn't a Google invention, but it's great to see the no-holds-barred Q&A ethos spread to larger companies. The same haircut bus that comes to Google on Monday goes to Yahoo on Tuesday. Dropbox had their first Take Your Parents to Work Day in 2013, and LinkedIn later declared November 7 to be theirs, with more than sixty employees bringing their parents into their New York office.[199] Maternity programs are improving across the industry. And on-site cafés are now a standard offering at Silicon Valley companies.

But the adopters of these programs still seem to be companies in the United States, in Silicon Valley. For anyone considering similar programs, I'll hazard a few reasons why you don't see a cornucopia of unique programs across different organizations. First, there's the false assumption that they cost money. It's just not true. Yes, in some cases there is opportunity cost (time spent on an ERG is time away from "work"), but as a practical matter that's more than returned in improved retention and happiness. Second, there may be a fear of entitlement. "If we invite a nail salon to come to our building to do manicures, won't employees be upset if we later eliminate the program?" That is a risk, but we tell employees up front when a program is a test, and that it will be continued only if it proves worthwhile. Third, managers may fear employees' rising expectations. If we do "this" today, then the workers will want "that" tomorrow. Again, honest conversation can address the issue before it arises. For example, when we were developing Google Shopping Express, a service where users can order from local stores and have products delivered to them the same day, we gave Googlers $25 to try it out. As the service developed, we'd periodically offer Googlers another $25 to spend. Each time, we explained it was a test, so there was never any expectation that each month you'd get a free $25. And when we stopped the tests, no one complained.

Fourth, and perhaps most important, it's hard to say yes. If an employee asks to bring an outside speaker to your office, think of the risks! The speaker may say something impolitic, it could be a waste of everyone's time, we don't have room, we're too busy, and most insidious, "What if I say yes and something goes wrong and I get in trouble?" It's very easy to find reasons to say no. But it's the wrong answer because it shuts down both employee voice and the chance to learn something new.

Find ways to say yes.

Employees will reward you by making your workplace more vibrant, fun, and productive.

One Googler, Gopi Kallayil, is in our sales organization but is also a Kirtan musician, practicing a kind of call-and-response chanting common in some spiritual traditions in India. He shared with me a CD of his music, and after I thanked him, wrote:

> You are welcome, Laszlo. Enjoy it and let me know if the music resonates with you and your practice. Last Monday we had live Kirtan music from a group of traveling international musicians called the Kirtaniyas. They played an acoustic set live in Charleston Park and my Monday yoglers [Googlers who practice yoga] group practiced yoga around them outdoors near the water cascade. The Googlers loved

it. Another great Optimize Your Life program organized organically by Googlers and at zero dollars cost. This is the secret to our culture. So much of it is grassroots-driven. The performers are in charge of the circus.

Be there when your people most need you

I must confess, though, that not everything we do falls neatly into our framework of efficiency, community, and innovation. Some programs exist purely because they make life better for our people. They simply felt like the right thing to do at the time we put them in place.

For example, one of the harshest but most reliable facts of life is that at some point half of us will be confronted with the death of our partner. It's a horrible, difficult time no matter what, made all the more tragic when it happens unexpectedly. Just about every company offers life insurance, but it never felt like enough. Every time we went through this as a company, we tried to find ways to help the surviving spouse of the Googler who'd passed away.

In 2011 we decided that if the unthinkable happened, the surviving partner should immediately receive the value of all the Googler's unvested stock. We also decided to continue paying 50 percent of the Googler's salary to the survivor for the next ten years. And if there were children, the family would receive an additional $1,000 each month until they turned nineteen, or twenty-three if they were full-time students.

We didn't tell anyone about the change, not even Googlers. It would have been a pretty macabre thing to advertise. We didn't think of it as a way to attract or retain people. There was no business benefit at all. It was just the right thing to do.

Eighteen months later, I realized I'd made a mistake. I was being interviewed by journalist Meghan Casserly for an article in *Forbes* magazine and I let slip that we offered this benefit. She immediately realized what a big deal this was, and published an article with the eye-catching title, "Here's What Happens to Google Employees When They Die."[200] The response was enormous and the article was quickly viewed almost half a million times.

Once word was out, I immediately heard from my peers at other companies. The number one question was, "Doesn't that cost a fortune?"

Not even close. Our cost thus far has been running at about one-tenth of one percent of payroll. Put another way, the average US company spends 4 percent a year on salary increases: roughly 3 percent for annual increases and 1 percent for promotions. I bet if you asked any employee if they'd want a program like this at the price of getting a 2.9 percent raise instead of a 3.0 percent raise, almost all would say yes.

In 2012 our benefits team received this anonymous email from a Googler:

> I'm a cancer survivor and every six months I have a scan to check if the cancer is back. You never really know when the news is going to be bad. But you inevitably start thinking about what you would do if it was bad. And so while I'm laying on that scanner bed I write and rewrite the email to Larry asking that my stock continue to vest for my family, even though I'm going to die.
>
> When I got your email about the new life insurance benefits it brought tears to my eyes. Not a day passes that I'm not appreciative of this company that does so many thoughtful and impactful things to my life. This life insurance is one of those things and it goes on the already long list of reasons I'm proud to work at Google.
>
> I had a scan two weeks after you announced the new benefits. I wrote no emails in my head.

I don't know who is responsible for this, but please pass along my deepest thanks to all those that were involved. You are doing amazing things to people's lives.

I'd erred in not announcing this benefit to Googlers. I hadn't thought of the stress and fear stalking Googlers such as the anonymous letter writer. Or of the impact we might have on other companies that would now explore this kind of program.

In 2011 we also changed our maternity leave policies. We decided to provide five months of maternity leave in the United States at a time when three months was more typical. But we made a more profound change as well. We decided that new parents would receive their full salary, bonus, and stock vesting for the entire time they were on leave. And any new parents receive a $500 bonus to help make life a bit easier by, for example, ordering home delivery of meals for the first few weeks.

By now, you'd guess that this was a data-driven decision, based on a detailed analysis of Googler happiness, retention, and the costs of the program—the kind of decision I've been advising you to make. But, like the death benefits, we made this call based on instinct. I was driving to work one day and thinking about the developmental differences between a three- and a five-month-old infant. I'm not an expert by any means, but I remember watching my own children grow. It is in the later months that babies start responding more to interaction, and new parents stop panicking at every cough and sneeze and realize that things are probably going to turn out okay. When I got to work, I said "Let's make the change."

Only later did I look at the data.

I learned that the attrition rate for women after childbirth was twice our average attrition rate. Many moms coming back to work after twelve weeks felt stressed, tired, and sometimes guilty. After making the change in leave, the difference in attrition rates vanished. And moms told us that they were often using the extra two months to transition slowly back to work, making them more effective and happier when the leave ended.

When we eventually did the math, it turned out this program cost nothing. The cost of having a mom out of the office for an extra couple of months was more than offset by the value of retaining her expertise and avoiding the cost of finding and training a new hire.

I was delighted to see that Facebook and Yahoo later followed in our footsteps, offering similar benefits to their new parents. I hope more companies do as well. You'll recall from chapter 2 the importance of mission, of your work being your calling. Few things drive that home better than receiving a message like Gopi's or Ann's, or an anonymous note of thanks. Imagine how work would feel for you if, instead of people coming to you with anxiety and desperation, they shared their gratitude for making their lives easier and for being there when they most needed support.

The best illustration I've ever seen of making a workplace special without breaking the bank comes from Michele Krantz, the principal of La Mesa Junior High School in Santa Clarita, California. Michele is the kind of leader who kicks off the week by standing at the front gate greeting each student by name and with a handshake. She walks around "with front-of-the-lunch-line passes in my pocket to give to kids being awesome." At her monthly staff meeting, her team presents Kudos granola bars to one another "to give them the opportunity to acknowledge each other." They literally give one another kudos. She writes birthday cards for every employee.

The result? At lunchtime students visit with and confide in her, faculty collaborate more than ever, and staff are committed. One of her groundskeepers, who had worked at the school for more than a decade, was so moved by her sending him a personal birthday card that he wrote her back. "He was so grateful," she told me, "and promised that he would work hard for me for the rest of his career."[201]

That wasn't her goal, of course. But it's a moving reminder of how the smallest investments of care and resources can have tremendous results.

- -

WORK RULES...FOR EFFICIENCY, COMMUNITY, AND INNOVATION

- ☐ Make life easier for employees.
- ☐ Find ways to say yes.
- ☐ The bad stuff in life happens rarely...be there for your people when it does.

- -

Nudge...a Lot

Small signals can cause large changes in
behavior. How one email can improve
productivity by 25 percent

The Temple of Apollo at Delphi, Greece.

The Greek geographer Pausanias, who lived from AD 110
to 180, visited the Temple of Apollo in Delphi. He saw a
stone in the forecourt inscribed with the Delphic maxim
gnothi seauton: Know thyself.

Sage advice, but hard to realize. We believe we know ourselves,
and that certainty is part of the problem. In his book *Thinking, Fast
and Slow*, Daniel Kahneman, a Nobel Prize–winning professor

emeritus of Princeton University, describes us as having two brains. One brain is slow, thoughtful, reflective, and data driven, and the other is fast, intuitive, and impulse driven. It's the second brain we rely on most, which is why even when we think we're being rational, we're probably not.

For example, how much is $5 worth to you? Would you be willing to leave a store and drive twenty minutes to save that much?

In 1981, Kahneman and his colleague, Amos Tversky,[202] wondered if we were consistent in how we valued money and time. They asked 181 people one of the following questions:

1. Imagine that you are about to buy a jacket for $125 and a calculator for $15. The calculator salesman tells you that the calculator you want is on sale for $10 at the other branch of the store, located a twenty-minute drive away. Would you make the trip to the other store?

2. Imagine that you are about to buy a jacket for $15 and a calculator for $125. The calculator salesman tells you that the calculator you want is on sale for $120 at the other branch of the store, located a twenty-minute drive away. Would you make the trip to the other store?

Keep in mind that these are 1981 dollars. If you adjust for inflation,[203] you'd almost triple the values. (Of course, you'd also have to adjust for the products, since no one under the age of twenty would know what a calculator was. "Let me get this straight. For $360 you want me to buy something bigger and heavier than my phone, and it won't even play Angry Birds?")

Tversky and Kahneman found that 68 percent of people "were willing to make an extra trip to save $5 on a $15 calculator; only 29 percent were willing to exert the same effort when the price of the calculator was $125."[204] Even though the savings were $5 in both cases, people would take action only when the $5 seemed large relative to the purchase price. The framing of the savings changed people's perception of value.

What about a more fundamental question? For example, how well do you think you can see something right in front of you? In *Sleights of Mind*, Stephen Macknik and Susana Martinez-Conde, laboratory directors at the Barrow Neurological Institute, describe how human vision is actually pretty awful, but we believe we see well because our brains fill in the gaps (and they go on to explain how magicians take advantage of that to deceive us). They suggest the following test:

> Separate out the face cards [from a deck of cards] and shuffle them. Fix your gaze on something directly across the room and don't let your eyes move at all. Draw a random face card and hold it out at arm's length at the very edge of your peripheral vision, then slowly pivot your arm forward, bringing the card toward the center of your unflinching forward gaze. Assuming you can resist the urge to let your eyes dart off to steal a glimpse, you will find that the card has to come quite close to your center of vision before you can identify it.

As they explain, "Your eyes can make out fine detail only in a keyhole-sized circle at the very center of your gaze covering one-tenth of one percent of your retina.... [Y]our vision is ninety-nine point nine percent garbage." But it doesn't feel that way because of saccades: rapid, intermittent movements of your eyes as they flit from one point to another. Your brain "edits out the motion blurs" and creates the illusion of continuous reality.[205]

If you still don't believe me, go to YouTube and type in "selective attention test." The first result should be a post by the University of Illinois's Daniel Simons. Go ahead. I'll wait.

...

...

...

Almost everyone misses what happens in the minute-long video (though I've already given you a hint that something funny might be going on). We believe we see what's happening around us, but we too often don't. Entire libraries have been written about the flaws in our wetware that not only skew our decision-making, but also keep us blissfully unaware of those flaws. Without realizing it, we are constantly nudged and buffeted by our environment, others, and even our own unconscious minds. Like deer wending through the woods by choosing the path of least resistance, we often rely on cues that operate below the conscious level to navigate our lives. When you drive on the highway, do you consciously choose your speed based on the posted limit and scrutinize your speedometer as you go? Or do you just go with the flow? When you buy clothes at a store, most salespeople hand you your purchase across the counter. High-end fashion retailer Nordstrom requires sales associates to walk out from behind the counter to hand you your purchase, which makes customers feel more personally cared for (and therefore more likely to keep shopping at Nordstrom). As a waiter, I used to squat down beside my guests' tables to talk to them. My being at eye level instead of looming above them made them more comfortable, and earned me larger tips.

In the fall of 2012, the Curve gallery at the Barbican Centre in London hosted an art installation called *Rain Room*, designed by rAndom International. The next summer, the installation moved to New York's Museum of Modern Art. *Rain Room* was an indoor, 100-square-meter (1,076-square-foot) field of falling water—it felt like it was raining indoors. As you walked through the room, the rain would stop falling around your body thanks to sensors that detected where you walked.

The *Rain Room* exhibit.

People waited as long as twelve hours to see the exhibit in both cities, yet London visitors spent an average of seven minutes inside while in New York—where the museum asked that people limit their visit to ten minutes and even gave a "courtesy tap" on the shoulder to those overstaying—many stayed forty-five minutes or longer. Both cities are similarly cosmopolitan; it doesn't seem that those in London would be any less interested in art or the rain; and the wait times weren't different. So what was?

The exhibit was free in London, but it cost $25 in New York.[206] Complementary to what we saw in chapter 7—where professors Deci and Ryan saw intrinsic motivation and productivity drop once they started paying people to perform tasks—once you charge for something, people think about it differently. They want to "get their money's worth." Without meaning to, and despite requesting that people limit the length of their visits, the Museum of Modern Art instituted an incentive system that caused exactly the behavior they hoped to avoid.[lix]

Even the physical layout of a place influences us in ways we don't realize. I remember visiting Hewlett-Packard's headquarters in 2011, a sea of taupe, high-walled cubicles stretching seemingly to the horizon. Hardly an environment that made it easy to ask a neighbor for advice: You couldn't even see your neighbor. In contrast, Mike Bloomberg, founder of Bloomberg L.P. and former mayor of New York City, arranged his own offices as a bullpen modeled on traditional newspaper newsrooms, optimized for swift exchange of ideas and information.

[lix] I'm assuming the Museum of Modern Art's goal was to have as many people see the exhibit as possible with a ten-minute-per-person limit. The London experience suggests they could have done that by waiving the admission fee. On the other hand, if their goal was mainly to maximize income, they didn't do too badly. The 55,000 visitors they had in the first forty-seven days of the exhibit paid $1,375,000 in admission fees alone.

Former New York mayor Michael Bloomberg in his office bullpen.[207]

Bloomberg sat in the very middle. Cubicles again, but what a difference. As a former employee told Chris Smith of *New York* magazine: "As a work space, it is something that you do not think that you can ever get used to.... But when you see the mayor hosting high-level meetings in clear sight of everyone else, you start to understand that this open-communication model is not bull****. And that it works"[208] [asterisks mine].

The common theme here is that we are far less consistent, objective, fair, and self-aware in how we navigate the world than we think we are. And because of this, organizations can help people make better decisions.

In their book *Nudge*, Richard Thaler and Cass Sunstein, professors at the University of Chicago and Harvard Law School, document at length how an awareness of the flaws in our brains can be used to improve our lives. They define a nudge as "any aspect of the choice architecture that alters people's behavior in a predictable way without forbidding any options or significantly changing their economic incentives....To count as a mere nudge, the intervention must be easy and cheap to avoid. Nudges are not mandates. Putting the fruit at eye level counts as a nudge. Banning junk food does not."[209]

In other words, nudges are about influencing choice, not dictating it. Some argue that nudges are unethical, forcing people into choices they would not otherwise make or want. But opponents of nudging ignore the reality that someone first made a choice *not* to put the fruit at eye level. Scottish philosopher David Hume described this problem, referred to as the "is-ought" fallacy or as Hume's Guillotine (because it severs the connection between "is" and "ought"). Just because something *is* done a certain way today, doesn't mean it *ought to be* done that way. In fact, many nudges are changes to poorly chosen current conditions that result in less health, wealth, or happiness.

For example, grocers place the most essential and perishable products like milk at the farthest end of the store, requiring us to traverse the well-stocked aisles at every visit. And they put high-margin, impulse-purchase items like candy bars at the checkout—even though you can purchase the same items elsewhere in the store—because that causes us to buy more candy bars. Would it be wrong or evil for them to instead put fruit by the registers? It would surely be less profitable, but it would just as surely be healthier for their shoppers. The purpose of grocery stores is not to improve their shoppers' welfare, however. It is to make money. Harsh, perhaps, but without profits a grocery ceases to exist. Even an all-organic neighborhood co-op grocer has to make enough revenue this month to pay salaries and rent, and buy inventory for next month. So I can't blame my local co-op for putting the three-dollar, organic, honey-sweetened, hand-churned peanut butter cup by the checkout. But I'd be healthier if I grabbed an apple to munch on instead of a candy bar.

Even if you accept that nudges can improve welfare, and you recognize that there's no reason to hold the status quo sacrosanct, there's still something we find unsavory about management's icy decision to manipulate employees through nudges. In a way, it's more comforting to know that the awful cube farms we live in are a result of simple ignorance or poor planning than to recognize that Bloomberg's bullpen was a calculated attempt to manipulate his teams into being more open and collaborative. How dare my facilities department/boss/government try to trick me!

On the other hand, aren't nudges just another tool in the management toolbox? One could argue that all management is an attempt to get people to be more productive, though I'll grant you that increasing happiness is not a universal management goal (although it should be—it works). Are nudges in the office completely different from a bonus plan that dictates how much a salesperson must sell?[210] Or from letting in more natural light because we believe that will make people more productive? Why would we feel more unsettled if we were told that the cube farms were deliberately designed and installed to isolate us from one another?[211]

I suspect there are two reasons why nudges give some people the heebie-jeebies. First, the specter of lab coat–clad brainiacs secretly influencing us is unsettling.

But nudges don't need to be secret. At Google we believe in transparency as one of our cultural cornerstones. We typically don't tell Googlers about our experiments as they are running, since that can change their behavior. After the experiment, however, we share what we found and how we intend to go forward.

Second, people don't like being reminded of the limits of our free will. Nudges raise all kinds of questions about desire ("Do I want a new Escalade because I need it, or because GM spent \$3,100,000,000[212] on advertising in 2012?"),[ix] about choice (Coke had 17 percent of the US market in 2012, compared to 9 percent

---

[ix] Advertising is of course Google's primary source of revenue, so as a shareholder I'm grateful for every one of those dollars!

for Pepsi,[213] even though one study using MRI scans showed that people found it hard to tell the difference),[214] and even about identity ("If the choices I make are a product of my environment and history, are any of my choices truly free?"). These are deep, deep questions, far beyond the scope of this book. But defensiveness is a natural reaction to being confronted with a threat to one's self-conception and identity.[215]

At Google we apply nudges to intervene at moments of decision in myriad ways. Most of our nudges are applications of academic research in a real-world setting. Jennifer Kurkoski, a PhD member of our PiLab team, jokes that "Too much academic research is just the study of college sophomores. So many experiments are conducted by professors offering five dollars to undergrads to participate in a study." Our approach is to take compelling academic findings, mix in some of our own ideas, and then see what happens when you try them on thousands of people going about their daily business. In doing so, and in writing about it, my hope is that others will benefit from our insights, whether you're a small business or a large one.

A sense of humanity compels us to be thoughtful, compassionate, and above all transparent when deploying nudges. The goal is not to supplant decision-making, but to replace thoughtlessly or poorly designed environments with structures that improve health and wealth without limiting freedom.

Our guiding principle? Nudges aren't shoves. Even the gentlest of reminders can make a difference. A nudge doesn't have to be expensive or elaborate. It only needs to be timely, relevant, and simple to put into action.

Much of our work in this area falls under a program called Optimize Your Life, which is led by Yvonne Agyei in partnership with Prasad Setty (People Operations), Dave Radcliffe (Workplace Services—he runs the shuttles and cafés, for example), and all of their teams. And Googlers, of course, are a great source of ideas and inspiration too.

Using nudges to help people become healthy, wealthy, and wise: wisdom first

We use nudges to make people happier and more effective. We saw earlier that something as simple as an email increased the rate at which women nominate themselves for promotion. And we're constantly looking for opportunities to provide advice or information at just the right time to improve how we work together.

Sometimes it's as simple as showing people facts. For example, one leadership team had developed a reputation for discord, with some members refusing to partner with one another and even undermining one another by withholding resources or information. "Performance management" didn't work because, regardless of their conduct, each person's overall output was quite good. "Coaching" didn't work because it was too time consuming, and two individuals in particular refused to admit any responsibility for the dynamic: "I'm not the issue," one told me. "It's the other people who aren't helping me!"

What did work was creating a quarterly survey of just two questions: "In the last quarter, this person helped me when I reached out to him/her"; and "In the last quarter, this person involved me when I could have been helpful to, or was impacted by, his/her team's work." Every member of the team rated each other member, and the anonymous ranking and results were shared with everyone. People knew where they fell in the ranking, but didn't know where anyone else fell. The two most obstreperous people, of course, ranked near the bottom, and were dismayed by it. Without any further intervention, they worked to improve the quality of their collaboration. Remarkably, in eight quarters the team went from 70 percent favorable on these questions to 90 percent.

Though not strictly a nudge, this was in line with work showing the power of social comparisons.[216] It was curious and heartening to see how simply presenting information and then relying on each person's nature—both competitive and altruistic—could transform a dysfunctional team. We had seen this effect in managers with our Upward Feedback Survey, but this was the first time we replicated it with a group of peers.

But what if we could get a team off to the right start and avoid its becoming dysfunctional? Then we wouldn't have to do any repair work. We decided to explore this idea by focusing on people who were both new to teams and new to the company: Nooglers.

A newly hired person actually destroys value. Imagine Ivan, a salesperson who earns $60,000 per year. Ivan costs $5,000 every month until he starts selling, and even once he's selling it takes time until his productivity exceeds his cost. He also consumes training resources and the time of the people around him whom he's pestering for advice.

And this isn't just a Google problem. Brad Smart, in *Topgrading*,[217] found that half of all senior hires fail within eighteen months of starting a new job. At the other end of the employment spectrum, consultant Autumn Krauss found that half of all hourly workers leave new jobs within 120 days.[218]

Making matters more difficult, Google managers were already busy and took a variety of approaches to acclimating new people to their teams, with no consensus about what worked best. The best case study we had for a long time was Kent Walker, who joined as our SVP and general counsel in 2006, whom we used as an example of a senior hire who had come up to speed and fully integrated himself into the company faster than almost anyone we'd seen. It had taken six months, compared to the full year it took most senior people, but Kent was unusually humble, curious, and self-aware.[lxi]

[lxi] Kent's secret: "I spent most of my time listening." Most people joined eager to get things done. But without understanding how to get things done *at Google*, they struggled. I used to call it "the Google crisis," a point about three or six months in when a new leader realized that the bottom-up, collaborative nature of our culture meant that they couldn't just bark orders and expect people to fall in line. We now build this learning into the first week of orientation for Nooglers.

We decided to prompt managers with a reminder about the very small tasks they could perform that would have the biggest impact on their Nooglers, and therefore the highest return on an investment of their valuable time.

In the pilot, managers received just-in-time emails the Sunday before a new hire started. Like the Project Oxygen checklist, which showcased the eight behaviors of successful managers, the five actions were almost embarrassing in their simplicity:

1. Have a role-and-responsibilities discussion.
2. Match your Noogler with a peer buddy.
3. Help your Noogler build a social network.
4. Set up onboarding check-ins once a month for your Noogler's first six months.
5. Encourage open dialogue.

And as with Project Oxygen, we saw a substantial improvement. Nooglers whose managers took action on this email became fully effective 25 percent faster than their peers, saving a full month of learning time. I was shocked at how profound the result was. How was it possible that a single email could have such a big effect?

It turns out checklists really do work, even when the list is almost patronizingly simple. We're human, and we sometimes forget the most basic things. Atul Gawande, whom we last saw in chapter 8, developed a surgical safety checklist that starts with "Patient has confirmed identity, site [i.e., where on the body the surgery is to take place], procedure, and consent."[219] The entire checklist has nineteen items. In 2007 and 2008, one hospital in each of eight countries[lxii] tried the checklist on a total of 7,728 patients. The rate of complications dropped to 7 percent from 11 percent. The rate

---

[lxii] The hospitals were in Amman, Jordan; Auckland, New Zealand; Ifakara, Tanzania; London, United Kingdom; Manila, Philippines; New Delhi, India; Seattle, United States; and Toronto, Canada.

of death dropped by almost half, to 0.8 percent from 1.5 percent.[220] Thanks to a checklist.

Granted, our stakes were considerably lower. No one physically dies of bad management (though perhaps one's soul does, a little bit). But simply sending out these five steps to your managers isn't enough. You have to send the checklist at the right time, make it meaningful, and make it easy to act on. We knew the email was timely, since it was going the night before someone started. We knew it was relevant, since the manager was probably wondering what the heck to do with this new person. Simple to put into action was trickier.

We first had to make sure the data was credible, so we included academic citations, results from internal studies, and the underlying data in the email. Googlers are, after all, data driven. We then had to make sure there were unambiguous steps the manager could take. Our people are smart but busy. It reduces cognitive load if we provide clear instructions rather than asking them to invent practices from scratch or internalize a new behavior, and this lowers the chance that an extra step might discourage them from taking action. Even the president of the United States limits the volume of things he needs to think about, so that he can focus on important issues, as he explained to Michael Lewis in *Vanity Fair*: " 'You'll see I wear only gray or blue suits,' [President Obama] said. 'I'm trying to pare down decisions. I don't want to make decisions about what I'm eating or wearing. Because I have too many other decisions to make.' He mentioned research that shows the simple act of making decisions degrades one's ability to make further decisions. It's why shopping is so exhausting. 'You need to focus your decision-making energy. You need to routinize yourself. You can't be going through the day distracted by trivia.' "[221]

So here's what we suggested our managers do for the first crucial interaction. I've left in the notation for footnotes and hyperlinks so you can see what Googlers actually receive:

1. Have a "role and responsibilities" discussion.

Research shows that having a clear understanding of one's job is associated with higher job satisfaction.[1] Here at Google, a study found that new grad hires who didn't understand their job expectations left Google in their first year 5x more often than those who did.[2] **What can you do?** Set up a meeting with your Noogler during his/her first week. Putting the meeting's agenda in writing is even better (see a template <u>here</u>). A few questions to answer for your Noogler: 1) What are Objectives and Key Results (OKRs) and what should your Noogler's first quarter OKRs be? 2) How does your Noogler's role connect with Google's business goals? The team's goals? 3) When will his/her first Perf [performance management] conversation be, and how will his/her rating be determined?

© Google, Inc.

At first blush all this specificity might seem a bit infantilizing, but in fact our managers told us it was liberating. Remember that not everyone is a born manager. By telling managers exactly what to do, we actually took one annoying item off their to-do list. They had less to think about and could focus instead on acting. The results recently moved a manager to shoot a quick thank-you note to the new-hire orientation team: "The operation you and your team have built is AWESOME! The email...with the onboarding information is emblematic of the entire experience. We really appreciate how easy you make this for us."

Back to the email: There's a lot going on in this initial "role and responsibilities" item. The first line cites the work of Dr. Talya Bauer,[222] a professor at Portland State University, who has done compelling research on what makes people happy and effective, and how it relates to your first experiences at a job. The second line shows that we see the same effect at Google, which in this case is huge and bears repeating: "New grad hires who didn't understand their job expectations left...5x more often than those who did."

We then outline very specific steps—a checklist inside a checklist (it's the turducken of checklists!)—followed by a link to a template and, for those too busy or lazy to click on the link, a few questions to get the manager started.

The remaining four checklist items followed a similar format. Significantly, they were all about helping the Noogler build a support network and developing a norm of clear communication. All the detail made the email a page and a half long, which we figured was the most we could expect someone to read. And it worked beautifully.

So that covered the managers of new hires, but what about the new hires themselves? Research on how people join teams and companies shows that some employees don't sit idly by, waiting for someone to "onboard them." Rather, they onboard themselves by reaching out to coworkers, seeking out resources, asking questions, and setting up lunches to build their networks. People who demonstrate this scrappy, proactive behavior become fully effective faster and perform better on tests of acculturation.[223]

As an experiment, we added a fifteen-minute segment to Noogler orientation for some people that explained the benefits of being proactive, provided five specific actions Nooglers could take to find the things they needed, and reiterated how this behavior fits with Google's entrepreneurial mindset:

1. Ask questions, lots of questions!
2. Schedule regular 1:1s with your manager.
3. Get to know your team.
4. Actively solicit feedback—don't wait for it!
5. Accept the challenge (i.e., take risks and don't be afraid to fail...other Googlers will support you).

Two weeks later, they received a follow-up email reminding them of the five actions.

Again, this doesn't look like rocket science, does it?

That's because when you design for your users, you focus on what is the minimal, most elegant product required to achieve the desired outcome. If you want people to change behavior, you don't give them a fifty-page academic paper or *ahem* a four-hundred-page book.

So where is the rocket science? It lies in identifying the general class of attribute that causes superior performance—in this case, proactivity—which comes from closely studying what the two tails are doing differently. In isolating the specific cluster of behaviors that exemplify proactivity for all Googlers to emulate. In developing a way to transmit those behaviors to those who wouldn't innately demonstrate them. And finally, in measuring the impact.

As we saw in chapter 5, we all think we are great at interviewing and assessing people, but by definition very few of us are truly great at it. The rest of us are just average. In almost every aspect of management practice and human relations, each of us believes we are in the top tail. And because we believe that, we continue designing ways to manage people based on gut instinct. The result is that we continue designing average-quality management systems that yield average results.

The good news is, we can learn to be better at it. We just have to pay attention.

So what were the results of this nudging experiment?

Nooglers who got the nudge were more likely to ask for feedback, became productive faster, and tended to have a more accurate sense of their own performance than a control group. But the subset of people who needed it most—those who were less naturally proactive—scored 15 points higher on measures of engaging in proactive behaviors during their first month than did others.

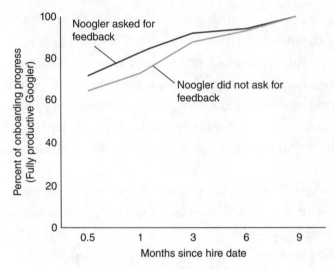

Percent of Nooglers who are fully productive.

The difference in the two lines adds up to a 2 percent improvement in productivity for the whole workforce. It's like getting a free employee for every fifty you hire, or in our case a hundred free employees for every five thousand we hire. Not a bad return for a fifteen-minute presentation and an email.[224]

There's an added benefit to giving Nooglers a nudge that complements what we do for managers. Even if the manager misses a checklist step, the Noogler picks it up. We're borrowing a concept called *poka-yoke* or "mistake proofing." It's a Japanese manufacturing concept, applied by Shigeo Shingo in his work at Toyota in the 1960s,[225] that shows up in many modern products. In most cars, failing to fasten your seatbelt causes an alarm to go off, signaling you that there is an error. The iPod Shuffle turns off automatically when you unplug your headphones, preserving the battery. A Cuisinart blender turns on only if the lid is securely locked, preserving your fingers. Similarly, we want to minimize the number of mistakes we make in getting Nooglers productive, and the best way is to put a reminder on both sides of the process.

The nudges to the Noogler and the manager, together with other changes we'd made to onboarding, allowed us to reduce the "time to become fully effective" to weeks instead of months.

Another challenge we had was that Googlers would often sign up for learning courses and then fail to attend. In the first half of 2012, we had a 30 percent no-show rate, which prevented Googlers on the waitlist from taking courses and caused us to run half-empty classes. We tried four different email nudges, ranging from appealing to our desire to avoid harming others (showing photos of people on the waitlist so enrollees could see who would be harmed by a no-show) to relying on identity consistency (reminding enrollees to be "Googley" and do the right thing). The nudges had the dual effect of reducing no-shows and increasing the rate at which people canceled their slots in advance, which allowed us to offer them to other Googlers. The effects of each nudge were different, however. Showing the photos of waitlisted enrollees increased attendance by 10 percent but did less to encourage unenrollment. Appeals to identity had the greatest effect on unenrollment, increasing it by 7 percent. Since then, we've incorporated these nudges into all course reminders and seen better attendance and shorter waitlists.

Nudges have proven helpful in changing norms at Google as well. Given how much information we share inside the company, controlling access to our buildings is critical. It's also not unheard-of in Silicon Valley for people to sneak into buildings to steal laptops and electronics, or even to try to get access to a company's systems. To prevent this, all external doors require the swipe of a badge for entry. But we also have a polite, friendly culture, and we found that Googlers were holding doors for one another because, after all, that's what our parents taught us to do. We sent emails asking people to check one another's employee badges (you typically clip them onto your belt) before holding doors for them, but it felt rude and awkward. People weren't really doing it. Until our security team put these stickers on every external door:

Googlers are nudged to be vigilant about on-campus safety thanks to this
sign on every building door.

This silly little cartoon, perhaps because it was so silly and
because it was posted at every door, broke the ice. It made it okay
for people to ask to see badges and check ID. As a result, the rate of
thefts and unauthorized entries dropped. Now when someone holds
the door for you, you'll notice that they always take a quick look at
your hips. Don't worry. They're just checking your badge.

Using nudges to help people become healthy, wealthy, and wise: becoming wealthier

Steven Venti and David Wise, professors at Dartmouth College and Harvard's Kennedy School of Government, penned a fascinating paper in 2000 that examined why households have different levels of wealth at retirement.[226]

Income is obviously a big factor. It makes sense that a family that has earned more money for thirty years will have more savings than a family that has earned less. Doctors typically end up with more savings than baristas, for example.

Venti and Wise sorted households into ten equal-sized groups called deciles, based on lifetime income as reported to the US Social Security Administration through 1992. The 10 percent of households with the lowest lifetime income were in the first decile, the second 10 percent were in the second decile, and so on, up to those in the top 10 percent of households who made up the tenth decile. Households in the fifth decile reported lifetime earnings of $741,587, twenty times higher than the first decile ($35,848) and less than half the tenth decile ($1,637,428).[227, lxiii]

But when they looked at the range of wealth within deciles, which allowed them to compare people with similar lifetime earnings, the results were shocking.[228]

Yes, households in the fifth decile average $741,587 in lifetime

---

[lxiii] Note that the income data is skewed low for several reasons. First, the sample is fifty-one- to sixty-one-year-olds, most of whom have years of earnings ahead of them (though their wealth will increase as well). Second, people who had zero income over the period (e.g., homemakers) are included. Third, income from transfer payments from the government is excluded. For example, about 10 percent of the US population received federally funded food benefits that year, which are not included here as earnings. Fourth, on the high end, lifetime earnings are understated, as only the first $55,500 of income was subject to social security tax at the time. Finally, the source data is from 1992. It doesn't change the findings, but to adjust to 2014 dollars you would factor all the values up by about 69 percent.

earnings. But their accumulated wealth—their savings, their invest-ments, their houses—ranged from roughly $15,000 to $450,000. In other words, when you hold earnings constant and look only at households with roughly the same lifetime income, the most wealthy have accumulated thirty times as much money as the least wealthy. And this general pattern persists at every income level. Look at the spread between the highest and lowest wealth levels for each decile. Even in the first (or lowest) decile, where substan-tially all income comes from government support, some households are able to accumulate $150,000 in wealth. This is a phenomenal achievement, requiring tremendous discipline at that income level.

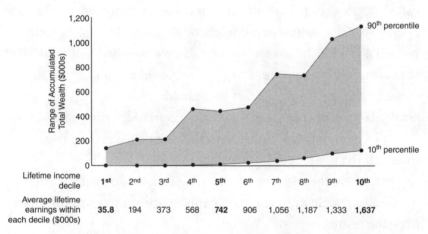

Relationship between lifetime income and wealth accumulation.

How is this possible? Are some households better investors? Do they have smaller families? Maybe a windfall inheritance? Is greater wealth due to some households being more risk-seeking, gambling on riskier investments that have greater payouts for some? Or per-haps families with less wealth had some unfortunate medical needs they needed to fund? Do they have caviar tastes?

Nope. None of these factors have a significant effect.

Rather, Venti and Wise explain, "most of the dispersion could be attributed to choice; some people save while young, others do not."[229] Stanford professor Douglas Bernheim and his colleagues examined the issues and came to the same conclusion. Households "differ in the extent to which they can exercise self-discipline over the urge to spend current income."[230]

I was skeptical when I read this because it seemed too obvious. The secret to being wealthy is just to save more money while you're young? And yet, even at the lowest income levels, some people were able to amass disproportionate wealth even when researchers controlled for the effects of the random financial triumphs and tragedies that beset us all.

Bernheim gives us a clue as to why: "If, for example, households follow heuristic rules of thumb to determine saving prior to retirement...then one would expect to observe the patterns [of wealth outcomes] documented in this paper." In other words, households tend to follow some gut-level rules about saving money, and those rules don't change.[lxiv]

Therefore, to improve savings rates early in life, which is the biggest determinant of wealth at retirement, households need to be somehow dislodged from their existing habits. In a forthcoming paper,[231] professors James Choi (Yale University) and Cade Massey (University of Pennsylvania), together with Jennifer Kurkoski of Google and Emily Haisley of Barclays Bank, argue that "a nontrivial portion of the variation in household wealth accumulation may be caused by variation in the savings cues individuals have been exposed

---

[lxiv] This probably deserves more than a footnote, but it's counterintuitive and stunning how much saving when you are young can affect your wealth. Let's imagine you start saving money at age twenty-five with a plan to retire at fifty-five, and earn an 8 percent annual return on your investments. To accumulate $150,000 over that period, you have to save only $110 per month. But you can get the same result by saving for just ten years if you can bump up your savings to $180 per month. On the other hand, if you wait until you're forty-five to start saving, you need to put away $460 per month to amass $150,000. The lesson is to start saving early, save as much as you can, and don't touch your savings. Compound interest is a powerful force. It's the difference between retiring well and not retiring at all.

to in their lives." Which means that very small differences in available information can significantly change people's behavior.

To test that idea, Jennifer partnered with Choi and Massey to run an experiment for us. The goal was to use small nudges to increase retirement wealth for Googlers.

In the United States, we offer a 401(k) retirement plan. In 2013, the Internal Revenue Code allowed employees to save up to $17,500 and delay paying taxes on the funds until they retire. Google matches 50 percent of employee contributions, adding up to $8,750 of free money to every Googler's retirement fund.

And yet not everyone participated. In fairness, not everyone had $17,500 to set aside. But even among those who did, participation was well below 100 percent.

The conventional wisdom is that this is because people have bills to pay, or would rather spend the money on discretionary goods, or simply because retirement is a long way away. But if the research is correct, it's not about any of these factors. It's just about being nudged to contribute at the right time.

In 2009, the more than five thousand Googlers who had not yet contributed the maximum and were not on track to do so received an email noting their year-to-date 401(k) contributions and one of four messages:

1. A basic reminder about the 401(k) program, which was our control for the study
2. The same as (1), but with an illustrative 1 percent increase
3. The same as (1), but with an illustrative 10 percent increase
4. The same as (1), but with a reminder that they could contribute up to 60 percent of their income in any pay period to help them catch up

What we didn't expect was that every one of these emails would trigger a reaction. Twenty-seven percent of Googlers receiving an email changed their contribution rates, and the average savings rate increased to 11.5 percent from 8.7 percent, regardless of which email someone received. Assuming an 8 percent annual return, this one year of contribution will add a total of $32 million to these Googlers' retirement funds. Assuming they continue at Google and keep contributing at that rate each year, every one of these Googlers will retire with an additional $262,000 in their 401(k)s. Even better, the people with the lowest savings rates increased the most, averaging a 60 percent higher increase than those in the control group. As one Googler wrote, "Thank you! I had no idea I was putting so little away!"

Each year since, we have continued to send out nudges, constantly tweaking them to further increase Googler savings. And each year, Googlers save more.

The nudge itself is cheap, though it results in greater expenditures for us as we continue to match retirement contributions. That said, it's money I'm delighted to spend.

Richard Thaler of the University of Chicago and UCLA professor Shlomo Benartzi conducted a series of experiments that were even more subtle.[232] They offered employees at three companies—an unnamed mid-size manufacturing company, a midwestern steel company called Ispat Inland, and two divisions of Philips Electronics—the opportunity to pre-allocate a portion of future salary increases toward retirement savings. Their Save More Tomorrow™[233] program had four features:

1. Employees were asked about increasing their retirement contribution rates well in advance of a scheduled pay increase.
2. The contribution change took effect in the first paycheck after a raise.
3. The contribution rate increased at each scheduled raise until it hit a preset maximum.
4. Employees could opt out at any time.

Over the next four salary-increase periods, 78 percent of employees who were offered the plan joined. Of those who joined, 80 percent stuck with it. And the average savings rate increased to 13.6 percent of salary from 3.5 percent over forty months.

It's a remarkable result, all the more so because the context is so very different from Google's. These are traditional companies with traditional workforces. It's inspiring and reassuring to see such a profound effect. These interventions together suggest that increasing savings, and better preparing individuals for retirement, is well within reach of most organizations. All it takes is a little nudge.

Using nudges to help people become healthy, wealthy, and wise: *Mens sana in corpore sano*[lxv]

In 2013, HR leader Todd Carlisle told a crowd at the Commonwealth Club in San Francisco that the ultimate recruiting slogan would be "Work at Google and live longer."

He wasn't joking.[234]

We'd been experimenting for years with ways to improve the quantity and quality of Googlers' lives as part of Optimize Your Life.

And because we provide free meals and snacks to Googlers, we found ourselves in a unique position to test whether insights from academic research would apply in the real world. Food is available from two sources: cafés, which generally serve two meals per day (either breakfast and lunch or lunch and dinner), and microkitchens, self-service stations stocked with drinks (soda, juice, tea, coffee, etc.) and snacks (fresh and dried fruit, crackers, chips, dark chocolate, candy, etc.).

Depending on the size of each office, we also offer on-site gyms, doctors, chiropractors, physical therapy, personal training, exercise/ yoga/dance classes, sports fields, and even bowling alleys. Medical services and personal training cost the same as outside Google,

---

[lxv] "A sound mind in a sound body," from the Roman poet Juvenal's *Satire X*.

but classes and facilities are free for everyone to use. We've experimented with many of these offerings, but here I'll just focus on some of the food-related results. Food is of course something we all have lots of experience with, but it's also a very simple, primal illustrator of how our impulses can override our conscious thoughts. Most of the insights gleaned from our food-related experiments translate directly to broader questions of how the physical space around us shapes our behavior, how many of our decisions are made unconsciously, and how small nudges can have large impacts.

Another reason I'm emphasizing food is that diet is one of the biggest controllable factors that affect health and longevity in the United States. More than one-third of adults are obese, which the Centers for Disease Control define as having a body mass index (BMI) of 30 or higher,[235] and require almost $150 billion in medical services each year.[236] Including those who are overweight (BMI between 25 and 29.9) brings the total to 69 percent of Americans. BMI is a measure of the relationship between your weight and your height. It's not a perfect measure by any means. For example, people with higher than average muscle development can have BMIs that suggest they are overweight, when in reality muscle is denser than other bodily tissues. But it's a simple-to-calculate starting point. BMI calculators are easy to find online if you want to check yours.

Managing your health, and your weight in particular, has all the hallmarks of an impossible task. The results are slow to appear and difficult to observe, so you get little positive feedback. It requires sustained willpower, which we all have in only limited supply.[237] And we're constantly bombarded by social pressure and messages that encourage us to consume more. Rob Rosiello, who ran McKinsey's Stamford, Connecticut, office while I was there, used to say that the most profitable line in the English language was "Would you like fries with that?"

This isn't a weight-loss book, and I'm far from an expert in health and nutrition, but the techniques we implemented at Google helped me lose thirty pounds over two years and keep it off. Even if you don't have cafés in your office, you may have a break room, or a vending machine, or a mini-fridge. And you've surely got a kitchen at home. Maybe some of what we've learned will help you too.

We decided to test three types of intervention:[lxvi] providing information so that people could make better food choices, limiting options to healthy choices, and nudging. Of the three, nudges were the most effective.

One study tested whether shocking signage would reduce consumption of sugary beverages. It was inspired by the work of Professor David Hammond of the University of Waterloo, Canada.[238] Starting in December 2000, Canadian cigarette packs were required to bear health warnings accompanied by graphic illustra-

"Shock signage" on a Canadian cigarette package.

Courtesy of Prof. David Hammond, PhD, University of Waterloo

[lxvi] Note that our experiments aren't perfect. Providing free food tends to cause overconsumption, at least until the novelty wears off. The constant and easy availability of food likely contributes to overconsumption as well. And our population isn't representative of the places we live. But we do apply the same rigor and tests of statistical validity as you would find in a peer-reviewed publication.

tions and blunt text about the risks of smoking, as can be seen in the side label. Hammond surveyed 432 smokers about the effect of the labels on their smoking habits over a three-month period. Nineteen percent reported smoking less as a result of the warnings. Those smokers who experienced fear (44 percent) or disgust (58 percent) in looking at the labels were much more likely to have reduced the amount they smoked and increased their likelihood of quitting.

We wondered if we could reduce the rate of sugary-beverage consumption with a similar approach, though of course soda and cigarettes are far from equivalent.

If you drink one can of soda
every week day for one year:

140 calories per can
x 260 week days
= 36,400 calories per year

3,500 extra calories
= 1 pound of body weight

You do the math!*

*But if you don't want to, that's 10 pounds per year

Sample shock signage used in our soda experiment. © Google, Inc.

We posted these signs in the microkitchens in one of our offices and monitored beverage consumption for two weeks before and after the signs went up. The signs had no significant impact, perhaps because they weren't startling enough, or the brand power of the soda overcame the shock of gaining ten pounds per year, or because the stakes were lower than with smoking.

We also tried color-coding food in our cafés, with red labels for unhealthy food and green labels for healthy food, which Googlers told us they appreciated but didn't lead to a measurable change in consumption. This is consistent with the findings of Julie Downs, an associate research professor at Carnegie Mellon, and Jessica Wisdom, a PhD member of our People Analytics team, who looked at whether publishing calorie information in two McDonald's locations in Manhattan and Brooklyn made any difference in consumption: It didn't.[239] Simply providing information wasn't enough to change behavior. Customers' buying habits didn't change at all, even once they could see that a ten-piece order of Mighty Wings had 960 calories, almost as much as two orders of large fries.[240]

If information by itself was insufficient, what if we reduced the range of options to include only healthy choices? I suspect it's this kind of approach that is feared by those who oppose nudging. Reducing choice also ran counter to our democratic impulses, but we wanted to be responsive to Googlers who were enthusiastic about getting people healthier, now.

Reducing options didn't work so well.

Our "Meatless Monday" pilot stopped serving land-based meat in two cafés on Mondays for one month. Attendance at one of the cafés dropped off, and the primary reason people gave for avoiding it was that they didn't like having choices made for them. As we'll discuss in the next chapter, there were much, much stronger reactions as well.

Googlers also told us they valued options. Fifty-eight percent of the comments across six microkitchen-based studies were supportive of more healthy food only if it was in addition to existing offerings. Googlers were willing to eat healthier, but not at the expense of choice.

Thus far we'd seen that people liked having more options and information, but no one was behaving differently. So we turned to nudging, subtly changing the structure of the environment without limiting choice.

The idea was sparked by an article by David Laibson, a professor of economics at Harvard University. In his paper "A Cue-

Theory of Consumption,"[241] he demonstrated mathematically that cues in our environment contribute to consumption. We certainly eat because we're hungry, but we also eat because it's lunchtime or because people around us are eating. What if we removed some of the cues that caused us to eat?

Rather than take away sweets, we put the healthier snacks on open counters and at eye and hand level to make them more accessible and appealing. And we moved the more indulgent snacks lower on our shelves and placed them in opaque containers.

In our Boulder, Colorado, office we measured the consumption of microkitchen snacks for two weeks to generate a baseline, and then put all the candy in opaque containers. Though the containers were labeled, it wasn't possible to see the brightly colored wrappers. Googlers, being normal people, prefer candy to fruit, but what would happen when we made the candy just a little less visible and harder to get to?

We were floored by the result. The proportion of total calories consumed from candy decreased by 30 percent and the proportion of fat consumed dropped by 40 percent as people opted for the more visible granola bars, chips, and fruit. Heartened by the result, we did the same thing in our New York office, home to over two thousand Googlers. Healthy snacks like dried fruit and nuts were put in glass containers, and sweets were hidden in colored containers. After seven weeks, Googlers in our New York office had eaten 3.1 million (3,100,000!) fewer calories—enough to avoid gaining a cumulative 885 pounds.

We turned to our cafés to see if a similar small nudge could change behavior. In a series of studies, Professors Brian Wansink (Cornell University) and Koert van Ittersum (Georgia Tech) demonstrated that serving-dish size has a powerful impact on food consumption.[242] They cleverly illustrated the issue by referencing the Delboeuf illusion, an invention of Belgian philosopher and mathematician Joseph Delboeuf in the late 1860s. The illusion looks like this:[243]

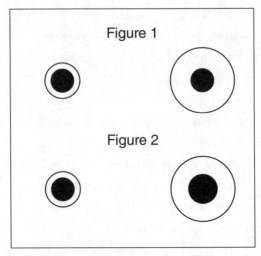

Illustration modeled on the Delboeuf illusion.

In figure 1, are the black circles the same size, or is one larger than the other? What about in figure 2?

In 1, even though the circle on the left looks larger, they are both the same size. In 2, even though the black circles look the same size, the one on the right is 20 percent larger. Now imagine the white circle is a plate, and the black is the food.

It turns out that in this case, seeing is believing. Our assessment of how much we eat and our satiation are heavily shaped by the size of the serving dish. The bigger the dish, the more we eat and the less satisfied we feel.

The professors presented six studies, one of which considered breakfast at a health and fitness camp. The subjects were overweight teenagers, who had been taught about portion control and how to monitor consumption. Experts, right?

Not even close. Campers who were given smaller cereal bowls not only consumed 16 percent less than campers who received larger bowls, they thought they consumed 8 percent more than the large-bowl campers. They ate less but were more satisfied, despite having been trained in how to measure and pace their consumption.

Another study was conducted at an all-you-can-eat Chinese buffet, where Wansink and professor Mitsuru Shimizu (Southern Illinois State University) observed diners who were given the option of choosing a small or large plate. Both sets of diners were comparable, with no difference in gender mix, estimated age, estimated BMI, or the number of trips to the buffet. At this point you won't be surprised to read that those who chose the bigger plate ate more. Fifty-two percent more, in fact. But they also wasted 135 percent more food, in part because they had larger plates that could hold more food, and because they left more of it on their plates.

The findings were intriguing enough that we wanted to try similar experiments at Google. Our primary goal was to improve Googlers' health, but since our cafés are effectively buffet restaurants we also thought we might reduce waste.

But the studies' samples were small and quite different from our workforce: 139 teenagers at camp and forty-three diners in the Chinese restaurant. Would the results hold for thousands of Googlers?

We selected one café and replaced our standard twelve-inch plates with smaller nine-inch plates. Sure enough, as we'd seen in other cases where we imposed a change without offering choice, Googlers got grumpy. "Now I have to get up twice to have lunch," wailed one.

Then we reintroduced choice, offering both large and small plates. The complaining stopped. And 21 percent of Googlers started using the small plates. Progress!

And then we added information. We put up posters and placed informational cards on the café tables, referencing the research that people who ate off smaller plates on average consumed fewer calories but felt equally satiated. The proportion of Googlers taking small plates increased to 32 percent, with only a small minority reporting that they needed to go back for seconds.

How did the sample size compare? We served over 3,500 lunches in that café that week and reduced total consumption by 5 percent and waste by 18 percent. Not a bad return on the cost of a few new dishes.

Design deliberately and with intent

Nudges are an incredibly powerful mechanism for improving teams and organizations. They are also ideally suited to experimentation, so can be tested on smaller populations to fine-tune their results. Prime Minister David Cameron of the United Kingdom established a "nudge unit" in 2010 that has improved collection of car taxes by 30 percent by sending overdue-tax notices along with a bold "Pay your tax or lose your [make of car]" label and a photo of the car. The unit also improved collection of court fines by 33 percent by sending notices by text message instead of mail. And in 2011 they replaced an attic insulation subsidy with an attic-clearing subsidy, with discounted rates for cleaning out clutter if insulation was subsequently installed. This cost people more, but the rate of adoption tripled.[244]

Up to 60 percent of people who are prescribed physical rehabilitation fail to complete their program or to fully comply with the rehabilitation protocols, according to Britton Brewer of Springfield College.[245] A Bay Area physical therapy center, PhysioFit, appears to see a substantially larger compliance rate. What do they do differently? They send a text message to all new patients: "Physio-Fit Physical Therapy and Wellness offers text reminders. Reply with 'Y' to sign up." The patient receives a text reminder for their appointments, as well as a daily reminder to perform their at-home therapy. Simple, free, and effective.

Richard Thaler reported on a change the state of Illinois had made to how it managed organ donor registrations.[246] In most states, as in Illinois before 2006, when you go to renew your driver's license you can check a box or fill out a form to indicate that you'd like to donate your organs, should you pass away and someone else be in need of them. About 38 percent of American drivers have taken this step to register as organ donors. Illinois made it just a bit easier. When you renew your license, you are asked, "Do you wish to become an organ donor?" Rather than waiting for you to sign up, they ask you. Three years later, Illinois had a 60 percent sign-up rate. Austria, where drivers have to opt out of organ donation, has a 99.98 percent participation rate. The same is true of France, Hungary, Poland, and Portugal.[247] As Hume argued, what *is* today may be different from what *ought* to be. And sometimes all it takes is a little nudge to get us there.

Ultimately, we are neither entirely rational nor entirely consistent. We're influenced by countless small signals that nudge us in one direction or another, often without any deep intent behind the nudges. Organizations make decisions about how to structure their workspaces, teams, and processes. Every one of these decisions nudges us to be open or closed, healthy or ill, happy or sad.

Whether you're part of a large organization or a small one, you may as well be thoughtful about the environment you create. Our goal is to nudge in a direction that Googlers would agree makes their lives better, not by taking away choice but by making it easier to make good choices.

WORK RULES... FOR NUDGING TOWARD HEALTH, WEALTH, AND HAPPINESS

☐ Recognize the difference between what is and what ought to be.
☐ Run lots of small experiments.
☐ Nudge, don't shove.

13

·························

It's Not All Rainbows and Unicorns

Google's biggest people mistakes and what you
can do to avoid them

Over the years, I've had dozens of skeptical executives tell me
that Google sounds too good to be true. No way can you
be completely open with your employees, because someone will leak your plans to the competition. If you give employees a
voice in running the company, they'll end up doing things management doesn't want. They never appreciate it when you do something
nice for them. It costs too much.

And there's some truth to those arguments.

Any idea carried to an extreme becomes foolishness. "Your right
to swing your arms ends just where the other man's nose begins,"
wrote lawyer Zechariah Chafee Jr. almost one hundred years ago.[248]
In assessing restrictions on free speech in the United States during
World War I, Chafee argued that "there are individual interests and
social interests, which must be balanced against each other, if they
conflict, in order to determine which interest shall be sacrificed
under the circumstances and which shall be protected."

One of the challenges of management at Google, indeed of
managing in any values-driven organization, is coming to a shared
understanding of "where the other man's nose begins." And it's in
the organizations with the strongest values that these distinctions
become most important. McKinsey & Company's website lists

"uphold the obligation to dissent" as a core value.[249] Marvin Bower, who became McKinsey's managing director in 1950, is credited with documenting and shaping the firm's values in a way that allowed McKinsey to serve its clients with integrity for more than fifty years.[250] For years, new associates were given an internally published copy of his book, *Perspective on McKinsey*. By the time I joined in 1999, *Perspective* was no longer being issued to new employees, though I found an old copy in someone's office. It detailed the early history of the company and the core beliefs that shaped the firm's ethics. Every associate bore the obligation to dissent, to speak up if they thought an idea was bad, wrong, or did a disservice to the client or the firm.

About a year later, I was serving a media client whose merger was yet to prove disastrous. The client had asked for a recommendation on how best to set up a venture capital business. The data were pretty clear. Aside from a few notable examples, like Intel Capital, most corporate venture-capital efforts were failures. They lacked the expertise, clarity of purpose, and physical proximity to where the most lucrative deals were being hatched. I told the senior partner it was a bad idea. I showed him the data. I explained that there were almost no examples of these kinds of efforts being successful, and none that I could find that were thousands of miles outside of Silicon Valley and run by people who lacked any engineering background.

He told me that the client was asking how to set it up, not whether to set it up, and that I should focus on answering the client's question.

Perhaps he was right. Or perhaps he had some superior insight into the issue that trumped my data. Maybe he'd already made that argument to the client, and they'd rejected it.

But to me it felt like I'd failed. I thought the obligation to dissent required me to speak up, so it was all the more gut-wrenching to see my concerns brushed aside. The more vigorously the firm proclaimed its values, the more acutely I and my peers felt the gaps between the values as they were espoused and as they were lived.

And it's not that McKinsey is a bad place. On the contrary, it's exceptional. I witnessed tremendous integrity, unyielding care for clients above all else, and a level of respect and collegiality accorded to people at every level that I still miss. It was a superb firm and training ground. But that moment stayed with me because in an environment with such a focus on values, even the slightest perceived compromises are felt disproportionately by people in the organization.

The same is true at Google.

We talk about values. A lot. And we're daily confronted with new situations that test those values. We are held accountable by employees, our users, our partners, and the world. We aspire to make the right decisions every time, but ultimately we're an aggregation of fifty thousand people. Sometimes some of those people make mistakes, and sometimes we as leaders make mistakes. We are far from perfect.

The test of the company, and of the management style I'm advocating in this book, is not whether it delivers perfection. It's whether we stay true to our values and continue to do the right thing even when tested. And whether we come through those challenges with a more refined commitment, shared among all Googlers, to our beliefs.

The price of transparency

I didn't have to wait long in my tenure at Google to see management's commitment to its values tested. My first TGIF started with Eric Schmidt onstage, pointing to a ten-foot-tall blueprint that was projected on the wall behind him. "These are the blueprints for the Google Mini," he announced. He showed us the inner workings of one of our first pieces of hardware: a product that you could buy, plug into your network, and instantly have your very own version of Google Search running on your company's intranet.

The room was filled with hundreds of Googlers, and thirty or forty Nooglers capping off their first week at Google. I was among the Nooglers, and none of us expected what came next.

"These blueprints were leaked outside of the company. We found the person who leaked them. And he has been fired." We believe we work better as a company if everyone can find out what everyone else is doing, Eric explained, which is why we share so much information among ourselves before any of it is publicly known. And it's why he was sharing the blueprints with everyone that day. He declared that he trusted all of us to keep information confidential. It was clear that if you violated that trust, you were gone the next day.

So the critics are right in principle but not in practice. We suffer about one major leak each year. Each time, there's an investigation, and each time, whether it was deliberate or accidental, well intentioned or not, the person is fired. We don't name the person, but we let everyone in the company know what was leaked and what the consequence was. Lots of people seeing lots of information inevitably means a few people screwing up. But it's worth it because the costs of leaks end up being small relative to that openness we all enjoy.

Reject entitlement

Entitlement, the creeping belief that just because you receive something you deserve it, is another risk in our approach. In a sense it's unavoidable. We are biologically and psychologically inclined to habituate to new experiences. People quickly become accustomed to what is being offered, and it becomes a baseline expectation rather than something wonderful and delightful. This can create a spiral of increasing expectations and decreasing happiness. It's part of why I like to bring new guests to Google, especially kids. It's embarrassingly easy to forget how unusual it is to have free gourmet meals served each day, but seeing a child's eyes light up when they realize they can get one of every dessert?

That never grows old.

When I joined Google, I had responsibility for the cafés, and it was a delight to work with gifted chefs like Quentin Topping, Marc Rasic, Scott Giambastiani, Brian Mattingly, and Jeff Freburg, along with Sue Wuthrich who led the café teams in addition to employee benefits.

But by 2010, the cafés had become an entitlement for a small but odious segment of Googlers. Instead of just eating at the office, some Googlers started packing food to take home. One afternoon I spotted a Googler putting four takeout containers in his car's trunk after lunch (which made me wonder: How healthy could it be to eat food that has been sitting in your hot trunk for six hours?). Another employee was caught filling his backpack with water bottles and granola bars on a Friday afternoon. He was going hiking on Saturday and wanted to bring enough food and drink for his friends. One Googler, outraged that we would dare to switch to smaller plates, wrote that she had started throwing out forks in some bizarre form of protest. Chefs shared that some Googlers had even thrown food at the café staff after being served.

But the final straw was Meatless Monday.

Meatless Monday, which I mentioned in the last chapter, is an initiative sponsored by the Monday Campaigns in association with the Johns Hopkins Bloomberg School of Public Health. It encourages people to avoid meat on Mondays as a way to improve their health and consume foods that require fewer resources to raise. For September of 2010, two of our twenty-plus cafés stopped serving land-based meat on Mondays, though they still served fish. We tracked café attendance on those days in September and compared it to August, and surveyed Googlers in the two Meatless Monday cafés, two control cafés, and online. Some people thought going vegetarian was great. Some didn't.

A vocal subset of Googlers in our Mountain View office were livid, sparking a lively set of discussions and holding a protest barbecue. The outrage was in part a reaction to more limited choice in those cafés, and in part a reaction to the perception that Google was mandating a certain ethos: that meat consumption was unhealthy.

There was nothing wrong with the protest barbecue. It was funny, clever, and a pointed critique. In a wry twist, at least one "protester" had asked a chef if they could not only borrow a barbecue but also have some meat to cook. And, of course, our headquarters in Mountain View, California, are just across the highway from an In-N-Out Burger and a host of restaurants, so there were ample meat-laden choices if you were willing to pay for your own lunch.

At the end of the month, we asked for feedback on Meatless Mondays. The complaints of the entitled few reached a crescendo, and I stood onstage at TGIF to share what we'd seen. I told Googlers about people ransacking the microkitchens, about the way our hardworking kitchen staff were being treated by a few petulant people, about the Googler who was throwing forks away. And I shared a piece of anonymous feedback that a Googler had written:

> Stop trying to tell me how to live my life. If you don't want to provide us the traditional food benefit then shut all the cafés down....Seriously stop this sh** or I'll go to Microsoft, Twitter or Facebook where they don't f*** with us.

The room froze.

The silent majority of Googlers, who had been unaware that this was happening, were appalled. They soon weighed in by the hundreds via email, at TGIF, and with private notes of thanks and support to our café team. Among them too were voices that wisely reminded us to assume good intent and be mindful of witch hunts. Even the person who was throwing away forks clarified that she of course hadn't meant that literally. Of course.

No one was fired, but the level of abuse and entitlement fell away. The mores had shifted.

A system like ours, which relies on people to be good and provides benefit of the doubt, is susceptible to bad actors. My very public statement, underscored by the appalling words of a fellow employee, made transparent what was happening and made it socially acceptable for Googlers to nudge one another. Snatching four takeout boxes at once was now greeted with a gentle "You must be pretty hungry," and people stockpiling snacks on Fridays were looked at askance.[lxvii]

Another way to address the challenge of habituation is by being unafraid to change benefits once the original reasons for them disappear. For example, in 2005 we started offering employees $5,000 in reimbursement if they purchased a hybrid car. The Toyota Prius had just been launched and was viewed as an experimental car, but we wanted to encourage employees to be environmentally responsible, and—because hybrid cars were allowed to use the High Occupancy Vehicle lanes on the freeway—we could reduce the impact our growing workforce was having on traffic in the community. A Prius cost about $5,000 more than comparable cars at the time.

Three years later, we announced in October that we were shutting down the program at the end of the year. The cars had become mainstream and prices had equalized—at this point we were just subsidizing Toyota. And we didn't have any evidence that this particular benefit was helping to either attract or retain people. Googlers were very upset at the change, because even after just a

---

[lxvii] The issue of entitlement isn't purely an internal issue for the technology industry, or for Google. Most of the people I've met in Silicon Valley are thoughtful, considerate people. But there are also real jerks—the kind of people whose windshields Wayne Rosing would have gladly smashed in—who would rather segregate themselves from our vibrant local communities. In San Francisco in particular, the influx of technology companies has driven up rents, and some argue that the tendency of these companies to provide meals and transportation for their employees has caused existing local businesses to lose foot traffic and customers. One start-up founder became a flashpoint for this frustration with his reprehensible "10 Things I Hate About You: San Francisco Edition" blog post, which fed just about every negative stereotype about the industry.

At Google we don't always get this right, but we try to be good neighbors through volunteerism, financial support of local schools and nonprofits, and buying from local growers as much as possible for our cafés. During the past several years we've donated $60 million to Bay Area nonprofits and volunteered hundreds of thousands of hours of Googler time.

We also focus on areas outside the limelight. For example, since 2013 we've partnered with the Berkeley County School District in South Carolina, near one of our data centers, to recruit and provide computer science and math teaching fellows to help more than 1,200 students become exposed to and interested in the fields.

few years they felt entitled to this program. Shutting it down served as a reminder that we provide our benefits and perquisites for specific reasons, and when the reasons are no longer valid we change our programs. At the same time, I softened the blow somewhat by explaining that we were also increasing our 401(k) contributions for employees.

But I do remember hearing reports from local dealers that Prius orders from Googlers tripled in December, just before the benefit evaporated.

"A foolish consistency is the hobgoblin of little minds"[251]

Each time we make changes to Google's performance management system, two truths become self-evident:

1. No one likes the system.
2. No one likes the proposed change to the current system.

We used to start our annual performance review process in December. It was a hearty undertaking: Everyone in the company received peer and manager feedback, performance ratings were assigned and calibrated across groups of managers, and it concluded with determining bonuses.

Our sales teams didn't like the timing. December was the last month of the last quarter of each year, when they were striving to finish deals. Other Googlers hated the timing too. They—quite reasonably—didn't want to spend time over the holidays writing reviews.

We asked ourselves, Why do we need to have reviews at year-end anyway? The timing didn't seem to work for anyone, and it smacked of "convention" in precisely the way that made Googlers want to reject it. We decided to move the annual reviews to March, well clear of the year-end crunch. I tested the idea with our management group, who, after some discussion, were supportive. The

People Operations team worked it through with their clients, who were also supportive.

On Thursday, June 21, 2007, at 5:04 p.m., I sent a note to our managers mailing list, which included thousands of managers, pre-announcing the change. I wrote that on Friday I would send an email to the company officially announcing the changes.

Aaaaand we had a "centithread," an email chain with at least a hundred responses.

Prior to my email, we had asked dozens of people, including top management, for input and had come to agreement.

Now we had thousands of people seeing the changes, and they absolutely did not agree.

By 6:00 p.m. I had given my (wonderful) assistant a list of forty people I needed to talk to that evening, based on the quality of their objections and how much their opinions influenced others. I called each one, listened, debated, attended to their core concerns, and tested alternatives. There were myriad reasons managers didn't like the proposal. Some people liked doing the bulk of their work earlier so they could relax over the holidays. Some people were busier at other times of the year and December wasn't an issue. Others preferred the convention of having a robust review before allocating bonuses, even if it meant substantially more work for them.

As I continued making calls and responding to emails, one particularly unhappy Googler emailed at 11:55 p.m. He wrote that what we were proposing was the opposite of what he thought engineers wanted.

It was almost midnight, and I asked if I could call him.

He said yes, and we spoke for thirty minutes.

What I'd learned throughout that evening was that the original proposal was wrong. Yes, management had agreed to it, and it had been represented to me that various constituencies had agreed to it. But it was wrong.

The next day, I emailed managers that rather than push reviews into March, we would start them in October. That both took pressure off the very end of the year and allowed detailed reviews to happen before bonus planning. Because there would be a gap between the end of reviews and bonus planning, we gave managers the freedom to adjust ratings in extreme cases if Googlers' performance trajectory changed materially.

Superficially, it looked like a vocal group of Googlers had shifted the result for the whole company. In reality, we ended up with a better solution precisely because Googlers were vocal. It wasn't easy to change direction publicly in front of thousands of people, but it was the right thing to do.

The experience underscored not just the importance of listening to our people, but also the need to have a reliable channel for opinions well before decisions are made. We eventually assembled a group we call the Canaries, engineers of varying seniorities, selected based on their ability to both represent the views of the various constituencies within engineering and credibly communicate how and why decisions were made. The group is named for the nineteenth-century mining practice of bringing canaries into coal mines to detect accumulations of toxic gases. Canaries, being more sensitive to the buildup of methane and carbon dioxide, would suffocate long before miners would. A dead canary meant it was time to evacuate. Similarly, our canaries give us an early warning about how engineers will react, and are trusted partners and advisors in shaping our people programs.

What struck me the most, however, was how much the people I called respected and appreciated my outreach. Jonathan Rosenberg once told me, "A crisis is an opportunity to have impact. Drop everything and deal with the crisis." Changing the timing of performance reviews is perhaps a trivial kind of crisis, but I dropped everything and spent the next eight hours making phone calls until well past midnight. We ended up with a better answer, shaped by the people who were most affected, and I gained a broader network of people I could turn to for counsel and help.

Treasure the weird

There have always been a handful of engineers who come to every TGIF. They sit in the front row and ask long, rambling questions. Every week. Newer employees sometimes roll their eyes when the same people get up every Friday and ask their questions.

More tenured people know better.

One of these questioners was a slight, brown-haired man. He had a gentle mien and always seemed to ask his questions in the form of a narrative. "Larry," he'd start, "I heard an interesting story recently that [five-minute digression]...and so I was just wondering if Google would [five-minute question]...?" The questions were sometimes wacky, sometimes prophetic. He asked about two-factor authentication[lxviii] years before it was offered.

Then one day, after a decade, he retired. The next week someone else was sitting in his front-row seat at TGIF.

It turns out he'd been one of our very early Googlers. I mentioned his departure to Eric Schmidt, who wondered if we weren't a bit poorer for having lost some of the quirky folks who'd been with us from the beginning.

We are.

Put more wood behind fewer arrows

Remember Google Lively? The product where you could create an animated avatar of yourself online and meet other people in simulated buildings and rooms?

Surely you remember Google Audio Ads, radio ads that were delivered by Google? Or Google Answers, where you could post

---

[lxviii] Two-factor authentication is a kind of security where, in addition to a password, you need a second piece of information to verify your identity. For example, when you buy gasoline you swipe your credit card and also need to enter your home zip code. Similarly, if you turn on two-factor authentication with Gmail, you'll need your password and also a numeric code generated by your phone or other device to log in.

a question, offering a bounty to anyone who answered it to your satisfaction?

These products were discontinued between 2006 and 2009. They were among the more than 250 products Google has launched over the past fifteen years, most of which not even I have heard of.

A side effect of untrammeled freedom is a flood of ideas. In addition to hundreds of products, we had a project database, where Googlers would log the thousands of 20-percent-time projects that had been started. We had an ideas board, where more than twenty thousand ideas were posted and discussed.

Despite the seething mass of activity, there was a sense that we didn't have enough people to do anything well. There were so many interesting projects going on that almost nothing had enough investment to be truly great.

In July 2011, our then SVP of Research and Systems Infrastructure, Bill Coughran, posted a blog entry titled "More Wood Behind Fewer Arrows." He explained that we were shutting down Google Labs,[252] a site where we allowed users to sign up for early trials of some of our products.

Behind the scenes, more was going on. Larry had pulled together the top two hundred or so leaders in the company and explained that we were trying to do too many things at the same time, and as a result weren't doing any of them as well as we could. He began leading an annual spring cleaning, shutting down products that weren't gaining traction (like Google Health, a site for storing your health information), were being done better by others (such as Knol, an attempt at an online encyclopedia), or had simply ceased to be relevant (Google Desktop, a product you could download to better search your computer's content, which became less relevant as most operating systems incorporated their own desktop search technologies).

None of these shutdowns were easy, as every product had fans, champions, and people working on them. Some wondered whether this new "top-down" focus on deciding which products live or die meant that our values had changed.

In fact, we'd rediscovered a principle we knew to be true in our earliest days: Innovation thrives on creativity and experimentation, but it also requires thoughtful pruning. With tens of thousands of employees and billions of users, there are infinite opportunities to create. And we attract people who want to do just that. But freedom is not absolute, and being part of a team, an organization, means that on some level you've agreed to give up some small measure of personal freedom in exchange for the promise of accomplishing more together than you could alone.

No single person could have created Google Search. Even at the beginning, it was Sergey together with Larry. And we have had ferocious arguments about how Search should operate. Over our history, we've replaced our search model several times, basically throwing out the old system (which brilliant, dedicated, creative people put thousands of hours into creating) to replace it with a better system.

The key to balancing individual freedom with overall direction is to be transparent. People need to understand the rationales behind each action that might otherwise be viewed as a step down the slippery slope that leads you away from your values. And the more central your values are to how you operate, the more you need to explain.

As important as explaining each decision is explaining the broader context. In October of 2013, a Googler asked me whether the annual product shutdowns were a sign that we cared less about individual ideas. I told him that the pendulum of letting one thousand flowers bloom[253], of fertilizing every idea, had swung too far and we were failing to make as much progress as our users deserved. Our product portfolio, like any garden, required regular, judicious pinching back. And in doing so, we would make our company healthier.

A most political dessert, or "You can't please all of the people all of the time"

At Google we have email lists to which one can subscribe, and a conversation related to a single topic is called a thread. Sometimes a particularly sensitive topic will spark a centithread. Our first millithread (or kilothread...there's some debate about the right prefix) had over a thousand responses. It was triggered by a pie.

One day in 2008, one of our cafés offered this dessert on the lunch menu:

> Free Tibet Goji-Chocolate Crème Pie with a Chocolate Macadamia Coconut Date Crust: Macadamia Crème, Cacao Powder, Vanilla Bean, Agave, Coconut Flakes, Goji Berries, Coconut Butter, Strawberry Infused Blue Agave Syrup, Medjool Dates, Sea Salt

Shortly after the menu was posted, a Googler emailed Eric and basically wrote: "This is from the menu today. If there is no good answer or action from the company I will quit in protest."

The Googler forwarded his note to several smaller mailing lists, and an engineer then re-forwarded it to the company-wide mailing list for miscellaneous topics.

It proceeded to set the record for the fastest time to get over a hundred responses, and became the first topic to break a thousand responses. One Googler counted the total at more than 1,300 emails about this topic.

And now, a bit of context.

The goji berry, also known as a wolfberry, originally comes from southeastern Europe and China and is now grown in Canada, the United States, and elsewhere.[254] It grows on bushes that can be three to nine feet high and bear purple flowers. The orangish-red berry itself is small, one to two centimeters long, high in antioxidants, and has a tangy, sweet-sour taste. I can't say I'm a huge fan, but they're not bad mixed into other dishes.

On that particular day in April, a chef decided to make a pie using goji berries that were from Tibet. All food at Google is free. So it was a free pie. With goji berries. From Tibet.

But for many Googlers, that was very different from a "Free Tibet goji-chocolate crème pie."

Google operates around the world, and we have several offices in mainland China. For many mainland Chinese, Tibet was, is, and should always be a part of China. For many others, Tibet was and should be an independent nation. Stepping back from Google, I'm probably not taking too much literary license to say that there are probably a billion people who hold one view, and a billion who hold the other view.

There were Googlers, like the one who emailed Eric and a few thousand of his closest friends, who were deeply offended by the implied notion in the pastry's name that Tibet should be "free." To make the point, some suggested that Westerners would be equally offended if "some chef in London" had offered "Free Wales pie" or a "Free Northern Ireland cookie." Others went so far as to suggest "Free Quebec maple syrup," "Texas polygamy steaks," and "War of Northern Aggression hotcakes."

One argument in response, made equally vocally by hundreds of people, was that this was simply freedom of speech. A chef should be free to name a dessert whatever he wanted. The debate then turned to whether speech within a company was truly free and, importantly, whether we were considering the beliefs and values of everyone in the company.

There was another line of debate about whether or not the chef should be punished. His manager had sent him home with a three-day suspension based on the initial Googler reaction, but many questioned whether that was fair, and indeed whether punishment would have a chilling effect on speech within the company. After all, they reasoned, if you got suspended for something like this, how comfortable would people be with any discussion? Had Google finally become a "big company" where it was verboten to say or think certain things?

And many Googlers thought the entire issue ridiculous. We were talking about the name of a pie. A pie. Really?

The debate didn't test just the limits of speech, but also how we comport ourselves when faced with intensely personal, emotional issues. As I read the hundreds of emails, I saw that there were fact-based arguments on both sides. I also saw that virtually no one was able to convince anyone else to change their views. People entered the debate believing Tibet was part of China or believing it wasn't, and exited the debate the same way. People thought this was either free speech or appalling insensitivity, and they thought so at the outset and at the end. Eventually, the rate at which people were commenting slowed and the thread limped to a close.

Nothing was resolved.

I realized that these kinds of enormous, brawling, wildly inconclusive debates are part of a culture of transparency and voice. Not every problem can be resolved with data. Reasonable people can look at the same set of facts and disagree, particularly where values are concerned. But the Googlers who wondered if penalizing the chef would chill speech at the company were on to something. The name of a dessert doesn't matter in the grand scheme of things. But if Googlers felt that they could be penalized for something as relatively trivial as this, then how could they ever be expected to ask our CEO the hard questions about whether we were staying true to our mission or putting users first? People knew they were free to have the conversation. And as painful as it is to live through, these debates are signs that you're getting something right.

Actually, one thing was resolved. I dug into the chef's suspension, and reversed it. He came back to work the next day. His intentions were good and no harm was done. His manager over-reacted, which was understandable in the face of hundreds of emails. I announced this in one of the last responses to the thread, and received more than two dozen notes thanking me, and management, for doing the right thing. The debate was important. And sparking a debate should never be a crime.

The leap of faith

Human beings are complicated, thorny, messy things. But those unquantifiable qualities are also what make magic happen. The purpose of this chapter is to lay bare some of the passion and pettiness that define our best and worst moments. Throughout this book I've tried to be honest about what has worked and what hasn't at Google, but I've leaned toward what has worked because that's a better roadmap for others.

At every step of the way, however, our deliberate choice to be mission driven, transparent, and empowering has brought along with it tension, frustration, and failure. At minimum, there's a chasm between the ideal we aspire to and the grungy day-to-day reality in which we all live. We will never be 100 percent transparent. No single Googler, not even Larry or Sergey, will ever have so much voice that they can dictate how every iota of the company operates, if only for the simple reason that if they tried to exercise that much control, people would quit. But unlike other environments I've seen, we recognize that our aspirations will always exceed our grasp. It's why achieving 70 percent of our OKRs each quarter is pretty good. And it's why Larry believes in moon shots, which cause you to achieve more in failure than you would in succeeding at a more modest goal.

Each one of the experiences I've shared here has made us stronger. We've refined our values, or at least reinforced the fact that we really do mean it when we talk about freedom.

Any team or organization trying to implement the ideas advanced in this book will stumble along the way, just as Google does. A few baby steps in, you will have your own "goji berry moment," where people get upset, or generate awful ideas, or take advantage of the organization's largesse. Not one of us is perfect, and a few of us are bad actors.

It is those moments of crisis that determine the future.

Some organizations will declare defeat, pointing to the smallest backsliding as evidence that people can't be trusted, that employees need rules and oversight to force them to serve the company. "We tried it this way," they'll declare, "and look where it got us. Employees got mad, or wasted money, or wasted my time."

Other leaders will prove to be made of sterner stuff. Those of you who, in the face of fear and failure, persevere and hold true to your principles, who interpose yourselves between the forces and faces buffeting the organization, will mold the soul of the institution with your words and deeds. And these will be the organizations that people will want to be a part of.

....

WORK RULES...FOR SCREWING UP

☐ Admit your mistake. Be transparent about it.
☐ Take counsel from all directions.
☐ Fix whatever broke.
☐ Find the moral in the mistake, and teach it.

....

14

....

What You Can Do Starting Tomorrow

Ten steps to transform your team and your workplace

My favorite video game of all time is called *Planescape: Torment*. Released in 1999, it opens with your character waking up in a mortuary with no memory. You spend the rest of the game adventuring across the universe only to discover (***spoiler alert***) that in prior lives you had done great good and great evil, waking after each life as a blank slate with the opportunity to choose once more how to live. At a key point in the game, you're confronted with the question, "What can change the nature of a man?" Your answer and conduct shape how the game unfolds.[lxix]

I wrote this book because, for better or worse, Google gets a lot of attention. In 2007 I was chatting with Dr. Larry Brilliant, who was leading Google.org, our philanthropic arm, and had decades earlier helped eradicate smallpox in India as a member of the World Health Organization. He repeated to me a comment Bill Gates had made, and though my memory is now a bit foggy, it was along the lines of "The Gates Foundation can give $100 million toward curing malaria, and get no attention. You launch a flu-tracking product and get media from around the world. It's not fair." For whatever

---

[lxix] You weren't expecting zombies, were you?

reason, people take a greater interest in what Google does than our size would suggest.

That level of attention carries with it responsibility. Our failures are more public, and since we're run by regular, fallible people, we have the same feet of clay as anyone else. We strive to apologize for and fix any mistakes. Our insights are also more public, and when we discover something, it creates an opportunity to share it with a much wider audience than we perhaps deserve.

Importantly, when we say things that aren't that revolutionary, that too gets attention.

In my years at Google and in writing this book, I noticed that many of the fundamental ideas we put into place were not that groundbreaking. But they still deserve attention.

You either believe people are fundamentally good or you don't.

If you do believe they're good, then as an entrepreneur, team member, team leader, manager, or CEO, you should act in a way that's consistent with your beliefs.

If people are good, they should be free.

Work is far less meaningful and pleasant than it needs to be because well-intentioned leaders don't believe, on a primal level, that people are good. Organizations build immense bureaucracies to control their people. These control structures are an admission that people can't be trusted. Or at best, they suggest that one's baser nature can be controlled and channeled by some enlightened figure with the wisdom to know what is best. That the nature of man is bad, and must be forged through rules, rewards, and punishments.

Jonathan Edwards, an American preacher who played a key role in the 1730s religious revival known as the Great Awakening, wrote a sermon that encapsulated this philosophy. It chilled me when I first read it in my high school literature class:

> Natural Men are held in the Hand of God over the Pit
> of Hell; they have deserved the fiery Pit, and are already

sentenced to it.... [T]he Fire pent up in their own Hearts is struggling to break out... [and] there are no Means within Reach that can be any Security to them.[255]

Now, the chilling of congregants was exactly what Edwards hoped to achieve. And as they say in government, mission accomplished.

Setting aside the religious context, which my tenth-grade class left me far from qualified to opine on, Edwards's underlying premise is that "natural Men" are bad and require some intervention to avoid a horrific end.

Steven Pinker, in *The Better Angels of Our Nature*, argues that the world has become a better place over time, at least when measured by incidences of violence. In the pre-state, hunter-gatherer era, 15 percent of people died violently, declining to 3 percent in the early Roman, British, and Islamic empires. By the twentieth century, homicide in European countries had dropped by another order of magnitude. Today, rates of violent death are even lower. Pinker explains that "human nature has always comprised inclinations toward violence and inclinations that counteract them—such as self-control, empathy, fairness and reason.... Violence has declined because historical circumstances have increasingly favored our better angels."[256] States expanded and consolidated, reducing the risk of tribal and regional conflict. People built bonds with one another through trade that made going to war ever more irrational. "Cosmopolitanism—the expansion of people's parochial little worlds through literacy, mobility, education, science, history, journalism and mass media...prompt[ed] people to take the perspective of people unlike themselves and to expand their circle of sympathy to embrace them."

Pinker lives in a world vastly different from that of Edwards. The world is more interconnected and interdependent. Yet our management practices remain mired in the mindset of Edwards and of Frederick Winslow Taylor, who told Congress in 1912 that management needs to tightly control workers, who were too feeble-minded to think for themselves:

> I can say, without the slightest hesitation, that the science of handling pig iron is so great that the man who is…physically able to handle pig iron and is sufficiently phlegmatic and stupid to choose this for his occupation is rarely able to comprehend the science of handling pig iron.[257]

Too many organizations and managers operate as if, absent some enlightened diktat, people are too benighted to make sound decisions and innovate.

The question is not what management system is required to change the nature of man, but rather what is required to change the nature of work.

In the introduction, I posited that there are two extreme models of how organizations should be run. The heart of this book is my belief that you can choose what type of organization you want to create, and I've shown you some of the tools to do so. The "low-freedom" extreme is the command-and-control organization, where employees are managed tightly, worked intensely, and discarded. The "high-freedom" extreme is based on liberty, where employees are treated with dignity and given a voice in how the company evolves.

Both models can be very profitable, but this book presumes that the most talented people on the planet will want to be part of a freedom-driven company. And freedom-driven companies, because they benefit from the best insight and passion of all their employees, are more resilient and better sustain success. Tony Hsieh of Zappos, Reed Hastings of Netflix, Jim Goodnight of SAS Institute, and many others will gladly tell you about the business results that come from giving their people freedom,[258] as the leaders of Wegmans and Brandix have. These technology companies have continued to see year after year of growth. Wegmans grows regardless of what happens in the economy, and continues to be a great place to work. What's beautiful here is that treating your people well is both a means to an end and an end in itself.

The good news is any team can be built around the principles that Google has used.

Throughout this book I've offered short lists of "work rules" in each chapter, in case you want to focus on one area or another. But if you want to become a high-freedom environment, here are the ten steps that will transform your team or workplace:

1. Give your work meaning.
2. Trust your people.
3. Hire only people who are better than you.
4. Don't confuse development with managing performance.
5. Focus on the two tails.
6. Be frugal and generous.
7. Pay unfairly.
8. Nudge.
9. Manage the rising expectations.
10. Enjoy! And then go back to No. 1 and start again.

1. Give your work meaning

Work consumes at least one-third of your life, and half your waking hours. It can and ought to be more than a means to an end. Non-profit organizations have long tapped meaning as a way of attracting and motivating people. For example, Emily Arnold-Fernández, founder of Asylum Access, a nonprofit that helps refugees, has built a world-class, global team based on a shared vision of enabling refugees to find work, send their children to school, and rebuild their lives in their new countries.

In too many environments, a job is just a paycheck. But as Adam Grant's work demonstrated, even a small connection to the people who benefit from your work not only improves productivity, it also makes people happier. And everyone wants their work to have purpose.

Connect it to an idea or value that transcends the day-to-day and that also honestly reflects what you are doing. Google organizes the world's information and makes it accessible and useful. Everyone who works here touches this mission, no matter how small the job. It draws people to the company and inspires them to stay, take risks, and perform at their highest levels.

If you're a lox slicer, you're feeding people. If you're a plumber, you're improving the quality of people's lives, keeping their homes clean and healthy. If you work on an assembly line, whatever you're making is used by someone and somehow helps them. Whatever you're doing, it matters to someone. And it should matter to you. As a manager, your job is to help your people find that meaning.

2. Trust your people

If you believe human beings are fundamentally good, act like it. Be transparent and honest with your people, and give them a voice in how things work.

It's fine to start small. Indeed, the less trust you have been showing, the more meaningful the smallest gestures will feel. For a company with a history of opaque management, a suggestion box that employees know is read and attended to will feel revolutionary. Let your teams ask you what motivated your latest decisions. If you're a small shop, regularly ask your employees what they would change to make things better, or what they would change if it was their company.

Because that's how you want them to behave. As if it were their company.

And the only way for that to happen is if you give up a little bit of your authority, giving them space to grow into it.

This may sound daunting, but in reality it's not too risky. Management can always take away the suggestion box, or tell people their ideas aren't welcome anymore, or even fire people. If you're worried that you might have to pull back your authority, let people know that each change is a test for a few months. If it works, you'll keep doing it. If not, you won't. Your people will appreciate even the attempt.

And if you are part of a team, make this plea to your boss: Give me a chance. Help me understand what your goals are, and let me figure out how to achieve them.

Small steps like these create the trail to an ethos of ownership.

3. Hire only people who are better than you

Organizations often act as if filling jobs quickly is more important than filling jobs with the best people. I've had salespeople tell me "Bad breath is better than no breath," meaning that they'd rather have the revenue that comes from a mediocre performer in a territory hitting 70 percent of their sales quota than have the territory empty.

But it is an error ever to compromise on hiring quality. A bad hire is toxic, not only destroying their own performance, but also dragging down the performance, morale, and energy of those around them. If being down a person means everyone else has to work harder in the short term, just remind them of the last jerk they had to work with.

Hire by committee, set objective standards in advance, never compromise, and periodically check if your new hires are better than your old ones.

The proof that you are hiring well is that nine out of ten new hires are better than you are.

If they're not, stop hiring until you find better people. You'll move more slowly in the short term, but you'll have a much stronger team in the end.

4. Don't confuse development with managing performance

Chris Argyris showed us how even the most successful people fail to learn. And if they can't learn, what hope is there for the rest of us? It's simply not pleasant to confront your own weaknesses. If you marry criticism with consequence, if people feel that a miss means that they will be hurt professionally or economically, they will argue instead of being open to learning and growing.

Make developmental conversations safe and productive by having them all the time, just like my manager used to do when we'd leave every meeting. Always start with an attitude of "How can I help you be more successful?" Otherwise, defenses go up and learning shuts down.

Development conversations can more safely happen along the way to achieving a goal. Separate in space and time conversations about whether a goal has been achieved. Once a performance period has ended, then have a direct discussion about the goals that were set and what was achieved, and how rewards are tied to performance. But that conversation should be entirely about outcomes, not about process. Goals were either missed, met, or exceeded, and each of those states carries different rewards or encouragements.

If you're doing this well, the performance discussions will never be a surprise because you'll have had conversations all along the way, and the employee will have felt your support at each step.

In all cases, don't rely solely on the manager to come up with an accurate picture of how people are doing. For development, solicit input from peers, even if it's as simple as asking or sending out a short questionnaire. For performance evaluation, require managers to sit together and calibrate their assessments as a group to guarantee fairness.

5. Focus on the two tails

Put your best people under a microscope. Through a combination of circumstance, skill, and grit they have figured out how to excel. Identify not just your best all-around athletes, but the best specialists. Don't find the best salesperson; find the person who sells best to new accounts of a certain size. Find the person who excels at hitting golf balls at night in the rain. The more specific you can be in slicing expertise, the easier it will be to study your stars and discern why they are more successful than others.

And then use them not just as exemplars for others by building checklists around what they do, but also as teachers. One of the best ways to learn a skill is to teach it. Enlisting stars as faculty, even if it's just for a thirty-minute coffee talk, will force them to articulate how they do what they do, and this very process helps them grow as well. If you're exposed to one of these people as a coworker, observe them closely, pepper them with questions, and use the opportunity to suck every bit of knowledge out of them.

At the same time, have compassion for your worst performers. If you're getting hiring right, most of those who struggle do so because you've put them in the wrong role, not because they are inept. Help them to learn or to find new roles.

But if that fails, exit them immediately. It's not mercy to keep them around—they'll be happier in an environment where they aren't the worst performers.

6. Be frugal and generous

Most things we do for our people cost nothing. Have vendors bring services in-house or negotiate lunch delivery from a local sandwich shop. TGIF and guest speakers require only a room and a microphone. Yet they result in a Brownian abundance: Googlers are constantly bumping into a new service or intriguing discussion.

Save your big checks for the times when your people are most in need, the moments of greatest tragedy and joy. Your generosity will have the most impact when someone needs emergency medical attention or when families are welcoming new members. Focusing on those most human moments underscores that your organization cares about each individual, personally. And everyone will draw comfort from knowing that should they encounter one of life's lowest or highest moments, they'll have the strength of a larger institution behind them.

This is true even for the smallest company. My father founded an engineering firm that he led for over three decades. He cared deeply for each of his people, paying them not just in wages but in kind words and advice and mentorship. And when any of his team reached five years in tenure, he took them aside for a private conversation. He told them that the company had a pension plan, and at five years they were fully vested in it. In addition to whatever they'd been saving, he had also been putting money aside for each of them. Some cheered, some cried, some simply thanked him. He didn't tell people earlier than that because he didn't want them to stay for the money. He wanted them to stay because they loved building things and loved the team. He was generous when it mattered most, and that made all the difference.

7. Pay unfairly

Remember that performance follows a power law distribution in most jobs, no matter what your HR department tells you. Ninety percent or more of the value on your teams comes from the top 10 percent. As a result, your best people are worth far more than your average people. They might be worth 50 percent more than your average people or fifty times more, but they are absolutely worth more. Make sure they feel it. Even if you don't have the financial resources to provide huge differences in pay, providing greater differences will mean something.

Your B players might be a little unhappy about their rewards, but you can address that by being honest: Explain to them why their pay is different and what they can do to change it.

At the same time, be generous in your public recognition. Celebrate the achievements of teams, and make a point of cheering failures where important lessons were learned.

8. Nudge

The single idea in this book with the most potential to tangibly improve the rest of your life is to change how much of each paycheck you save.

If you compare people who have earned the same amount over

thirty years, there can be a 3,000 percent difference in how much wealth they accumulate, and it is almost entirely determined by how much they save. And it is almost never easy to save. Unless you are richer than Croesus, every dollar of savings will feel like a trade-off. Do I buy the name brand or the generic? The $3 peanut butter cup or skip dessert? Trade in your car or hold on to it for one more year? In my first year after college, while working as an actor and a waiter, I used to frequent the Hostess Thrift Shop near my town, which sold bread and pastries that were just at the edge of expiration. I got my snack cakes (in moderation!), and was able to save a few extra dollars each week. Remember that getting Googlers to tweak their savings rates by less than 3 percent will add $262,000 to each Googler's retirement fund.

As crazy as this sounds to many, I know of people who view their $100,000 summer rentals in the Hamptons as a necessity, and even when my banker friends lost their jobs in 2008, they found a way to escape to their beach houses.

I'm belaboring the point, but people tend not to change their savings rate. Figure out what percent of your income you save today, and then save a little bit more from now on. It is never easy. It is always worth it.

That's for you personally.

Now look around you right now and discover how your environment is nudging you and those around you already. Is it easy to see other people and connect? Are the least healthy snacks in your refrigerator at eye level? When you email or text your colleagues and friends, is it to share good news or snark? We are all constantly nudged by our environment and nudging those around us. Use that fact to make yourself and your teams happier and more productive.

Arrange your physical space in a way that encourages behaviors you want: If you need people to collaborate and are stuck with cubicles, knock down the walls. Be thoughtful in how you send messages to your teams. Share data about what is going right, such as the number of people volunteering with local charities, to encourage others to get involved. You'll be surprised by how different the same place can feel.

9. Manage the rising expectations

You'll trip up sometimes and need to take backward steps. Be prepared to eat your goji pie. Knowing that, tell people around you that you'll be experimenting with ideas from this book before you start experimenting. That will help transform them from critics to supporters, and they'll extend you more benefit of the doubt if things go awry.

10. Enjoy! And then go back to No. 1 and start again

Larry and Sergey set out to create the kind of place where they'd both want to work. You can do the same thing. Even if you join a company fresh out of school, as a junior employee, or as employee number 1,000,006, you can still be a founder by choosing how you interact with those around you, how you design your workspace, and how you lead. In doing so, you'll help create a place that will attract the most talented people on the planet.

This isn't a one-time effort. Building a great culture and environment requires constant learning and renewal. Don't worry about trying to do everything at once. Experiment with one idea from this book or with a dozen, learn from the experiment, tweak the program, and try again.

What's beautiful about this approach is that a great environment is a self-reinforcing one: All of these efforts support one another, and together create an organization that is creative, fun, hardworking, and highly productive.

If you believe people are good, then live your beliefs through your work.

Google has been named the best place to work over thirty times by the Great Place to Work Institute and has received hundreds of other honors from organizations that champion women, African

Americans, veterans, and others, and from governments and civic organizations. But we're not the first "best place," nor will we be the last, nor are we even the only one today.

One thing Google does specialize in is operating at scale, building systems that serve two billion users just as thoughtfully and reliably as they serve ten. Our people innovations have benefited from a set of prescient founders, fierce cultural guardians, thoughtful academic research, and creative companies and governments. Thousands of Googlers have shaped how we choose to operate, pushed us to find ever more creative and equitable solutions to people problems, and held us accountable. And I've had the humbling experience of being surrounded by colleagues and a People Operations team so insightful, committed, and creative that I struggle to keep up. I draw inspiration from them every day.

Tens of thousands of people visit our campuses each year and ask, "Why are people here so happy?" "What is Google's secret?" "What can I do in my organization to make it more innovative?"

The answer is in your hands.

....

WORK RULES

1. Give your work meaning.
2. Trust your people.
3. Hire only people who are better than you.
4. Don't confuse development with managing performance.
5. Focus on the two tails.
6. Be frugal and generous.
7. Pay unfairly.
8. Nudge.
9. Manage the rising expectations.
10. Enjoy! And then go back to No. 1 and start again.

....

Afterword for HR Geeks Only: Building the World's First People Operations Team

The blueprint for a new kind of HR

Some readers will be interested in how this all happens at Google. While the underlying impetus for how Google treats people emanates from the founders, it falls to the People Operations team to ensure we are living up to our aspirations, and exceeding them.

Until 2006, this team was called Human Resources and the job I was recruited for was titled Vice President of Human Resources. But when I received my offer letter, the job title instead was Vice President of People Operations. As crazy as this sounds today, I wasn't thrilled by this turn of events. Executives tend to fail in their jobs about a third of the time, and I was about to move my young family from New York to California to join an organization that my division CEO at GE had called "a cute little company." I worried that an oddball title like People Operations would make it that much harder to find another job if things didn't work out.

I called Shona Brown, then Google's SVP of Business Operations (and herself a former McKinsey partner and Rhodes scholar), and asked if I could have the original title. I confess only now that I didn't tell her why.

Shona explained that at Google, conventional business language wasn't well regarded. "HR" would be viewed as administrative and bureaucratic. In contrast, "operations" was viewed by engineers as a credible title, connoting some actual ability to get things done. And, importantly for an HR job, "operations" would also suggest that I could actually do math!

Shona and I agreed that I'd start with the "People Operations" title, and after six months could revert to "HR" if I chose.

When I joined, I met with each of Google's twelve top leaders one-on-one to introduce myself and understand their needs. Urs Hölzle, whom we met in chapter 4, was the senior vice president of infrastructure at the time and one of the first ten people hired by Google. He had been a brilliant professor of computer science and had already founded and sold one start-up, Animorphic Systems. Urs was lured from teaching to design and build Google's data centers—no mean feat, given that Google backs up the Internet many times over!

In our first meeting, Urs shook my hand, looked at my resume, and said: "Great title."

I never changed it.

Since then, we've built People Operations around four underlying principles:

1. Strive for nirvana.
2. Use data to predict and shape the future.
3. Improve relentlessly.
4. Field an unconventional team.

Strive for nirvana

For HR professionals reading this book, much of what we do at Google may sound like nirvana, a near-unobtainable ideal, but it started very simply. I went into my first one-on-one meeting with Eric Schmidt with grandiose ideas about the kinds of programs we could develop to better manage careers and help senior leaders develop. Eric wasn't particularly interested in my strategic vision. He had more pressing concerns.

Google had almost doubled from about three thousand people in 2004 to 5,700 people in 2005. Over the next year, Eric knew, we were going to do it again, ballooning to almost 10,700 people. We needed to go from hiring fifty people each week to almost a hundred, without compromising quality. This was the biggest people challenge we had.

I'd made an amateur mistake. Before Eric would entertain any esoteric ideas, People Operations had to deliver on Google's most important issue. I learned that to have the privilege of working on the cool, futuristic stuff, you had to earn the confidence of the organization. In 2010, we enshrined that notion in a graphic that encapsulated our approach. The pyramid shape was a nod to psychologist Abraham Maslow's hierarchy of needs,[259] which he presented as a pyramid with our most fundamental needs at the bottom (air, food, water), topped by the need for safety, belonging, love, and finally self-actualization. After seeing our version, a few folks from our team took to calling it Laszlo's hierarchy.

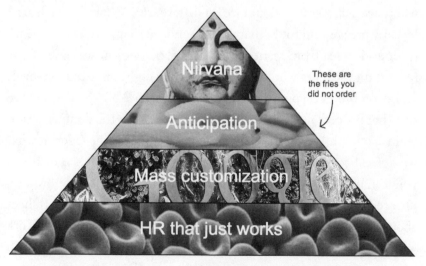

"Laszlo's hierarchy." © Google, Inc. Image by Elizabeth Yepsen.

This was our path to achieving HR nirvana, that blissful place where every Googler was growing seemingly without effort, as our programs worked behind the scenes to fill every job, create learning opportunities, and help Googlers be more productive, healthier, and happier.

I picked the graphic of blood cells to illustrate "HR that just works" to underscore the point that our programs are as ubiquitous, and need to be as reliable, as the body's circulatory system. We must deliver the basics, flawlessly, every time. No errors in offer letters or bonuses, every job filled on time with a great candidate, smooth and fair promotion processes, speedy resolution of employee concerns, and so on. This level of consistent, high quality in all our operations was how we would earn the right to do more. Whatever your own aspirations may be, this is the starting point. Otherwise, failure to deliver the basics even once in a while will cause your business to withhold trust and authority when you want to take on more.

Our compensation team, for better and worse, always received tremendous attention from the management team. To make sure everything works and we stay ahead of management's expectations, there's a formal debrief after every process such as bonus planning, where we ask, "What should we do differently? What did we learn? What were we told to do that we will choose to ignore and not do?" (Not every idea from management is a good one—it was one of our most senior leaders who suggested we have eight hundred job levels and promote everyone four times each year!) And then, when we run that process next time, the first conversation the compensation team has with Google's leadership team starts with, "Here's what we agreed to last time, and here's what we did. Here's the things you told us to do that we will not do, and here's why. And now, let's proceed." The compensation team even developed a cheat sheet about each management team member, describing how best to work with him or her, so that new team members can work smoothly with our most senior leaders right from the start. (I'd love to share some examples with you, but they apparently have a cheat sheet about me as well and refuse to show me any of them.)

Mass customization, the second step on the path to nirvana, was a departure from our past approach. The notion of mass customization comes from author Stan Davis, who wrote a book called *Future Perfect* in 1987, describing a world where companies will produce goods and services to meet individual customers' needs with near–mass production efficiency. And that's what we try to do. The visual metaphor was a forest: every plant unique in size and shape, but nevertheless having more in common with the others than not.

We've always had a consistent philosophy underpinning our people processes, but tweak the details of each process based on what different parts of Google need. Years before, we used to require that people processes, such as promotion decisions and even whether or not ratings were disclosed to employees, all follow the same protocol. In fact, Tiffany Wu, an analyst on our team, used to maintain a compliance checklist posted on her wall, with good and bad marks for each VP, tracking whether they had dutifully disclosed ratings to each Googler or hit a prescribed distribution of salary increases. As we grew, enforcing such extreme homogeneity stopped making sense because some groups are authentically different from one another. While our best engineers may have hundreds of times more impact than an average engineer, our best recruiter doesn't have hundreds of times more impact than average. It didn't make sense to enforce the same kind of reward distribution on both populations.

Or consider our engineering promotion process, where potential promotions are reviewed by committee and ratified by a subsequent committee. If a Googler disagrees with the decision, there is an appeals committee, and if that decision isn't well received there's an appeal-of-appeals committee. When I described it to Google board member John Doerr, the managing director of venture capital firm Kleiner Perkins Caufield & Byers, he said, "Even as an engineer, I'm amazed that someone would design a process this byzantine." But it works because these checks and balances ensure that processes are just, and that they're as transparent as possible— qualities that matter deeply to engineers. In some sales teams, leaders may instead say, "You know what? We're just going to make a call and the decision will be final," and there's not an appeals process. That's viable as well, because in those cases People Operations works behind the scenes to ensure the processes are just as fair by enforcing the same talent standard across the company. Same standards behind the scenes but different manifestations on the front end, where Googlers see them. And then, in the spirit of transparency, we share data on the outcomes of each promotion process with Googlers, along with historic data.

In most HR departments, there's a bias toward consistency as a mechanism to ensure fairness, but our friend Emerson would remind us that there's a difference between consistency and foolish consistency. For example, at GE, granting special bonuses above a certain amount ($5,000 or so, if I recall correctly) required approval of the CEO, Jeff Immelt. That may have made perfect sense for the industrial side of the company, where only executives were eligible for bonuses and granting one to a nonexecutive was a truly exceptional event. But at GE Capital, the financial services side, bonuses were much more common, mirroring practices in that industry. Applying the same approval standard to Capital frustrated managers, and made the HR people tasked with upholding the policy look like petty bureaucrats. If you're an HR practitioner, you must constantly ask yourself whether the principle underpinning each rule is relevant to the case at hand, and fearlessly abandon practice and policy when the situation merits it.

The french-fry graphic labeled "anticipation" needs a little explaining. I drew the name from an episode of the comedy *30 Rock*. Set at NBC's headquarters in Rockefeller Center, the show followed the cast and crew of a variety show, starring comedian Tracy Jordan (played by the real-life comedian Tracy Morgan). In one episode, Tracy is furious because his staff has brought him a hamburger, but failed to bring the fries he did *not* ask for: "Where are the french fries I didn't order?! When will you learn to anticipate me?!"

When I first saw the episode, I thought Tracy was a hilarious monster of ego.

Then I realized he was right. He wasn't a psycho. He was an executive!

People are happy when you give them what they ask for. People are delighted when you anticipate what they didn't think to ask for. It's proof that they're wholly visible to you as people, not just as workers from whom you're trying to squeeze productivity.

Anticipation is about delivering what people need before they know to ask for it. Thanks to *30 Rock*, we call these instances of perfect anticipation "french fry moments."

Take the $500 we give each Googler after childbirth to spend on home delivery of meals. The first few days and weeks after bringing a new child home are exhausting. The last thing anyone wants to do is cook. Even though Googlers can afford to order a pizza for dinner, the mental accounting is different when someone gives you $500 specifically for takeout meals. And new parents tell us they love it.

Ironically, our first executive development programs like the ones I had proposed to Eric in our initial meeting turned out to be a classic french fry moment. When Evan Wittenberg (then a member of our learning team and now SVP of People at Box, an online data-storage company), Paul Russell (an early leader of our learning team, now retired), and Karen May (at the time a consultant and our current VP of People Development) created Google's first Advanced Leadership Lab in 2007, it was intensely controversial because Google at the time was organized by function—engineering, sales, finance, legal, for example—and the groups didn't interact unless necessary. Most leaders knew who the key people were in each function and could reach out if they had to. Googlers didn't see the need to bring people together from different functions into a training program, and certainly didn't see the need to distract our leaders from their important work by dragging them off-site for three days. But by the time we hit twenty thousand employees at the end of 2008, it was no longer possible for our leaders all to know one another, and the connections forged at the labs became vital. Evan, Paul, and Karen anticipated the need for this program almost two years before we needed it, and thus had time to fine-tune it so that by the time it was essential, Googlers were telling us it was one of the most crucial and effective development programs they'd ever experienced.

As you think about creating your own french fry moments, keep in mind that they are thankless. You rarely get praised for avoiding a problem. It's why in politics you can never win points by arguing "But the recession would have been so much worse if not for my policies!" But you'll know, and your team will know. And your company will run better. And people will be happier.

Once you've scaled this pyramid, which isn't that different from Maslow's, you achieve HR nirvana. For employees, it means that you enjoy what feels like a random walk through Google: You have a terrific set of interviews with compelling people, you join and feel welcome, becoming productive in a few weeks because you've met helpful people, and you're constantly surprised as opportunities unfold ahead of you. It's like one of those choose-your-own-adventure books that some of you may have read as kids, where every page opens more and more options. Along the way, you're becoming a better leader and a better entrepreneur. That's what HR nirvana looks like if you're a Googler. And behind the scenes, People Operations aspires to have thought through the experience meticulously, clearing away all the stones in your path so there's nothing to trip you up.

Use data to predict and shape the future

If you've read this far, you will be shocked, shocked to find that working with data is central to how we built and run People Operations. But we started very small. Our Analytics group was started when I asked three analysts who sat in different groups (staffing, benefits, and operations) to join forces and compare notes. They were initially resistant, having little interest in one another's areas. But soon the operations analyst was teaching the others how to program, and the staffing analyst was teaching the others advanced statistical techniques. And together they laid the foundation for our analytics today.

Prasad Setty describes the evolution of analytics as moving from description to analysis and insight to prediction, using employee attrition as an illustration.

In most companies it's distressingly difficult to report on seemingly easy questions like, Who has given notice to resign but hasn't hit their last day yet? How many employees do we have? Or even, Where is everyone? Employee data is housed in multiple computer systems that update with different frequencies and may not talk to one another. The payroll system needs to know where people are located for tax purposes, but that may be different from where they are working—for instance, if a British employee is spending two months in New York on an assignment. Even definitions of basic concepts like "current employees" often vary from department to department. Finance may count anyone who is employed more than one hour per week as a current employee, but the benefits department likely only includes people who work more than half-time and are therefore eligible for benefits. The staffing team, meanwhile, might include people who have accepted offers but not yet started, since that gives a better sense of how close they are to achieving their hiring goals.

So step one, and it's a big step, is agreeing on a common set of definitions for all people data. Only then can you accurately describe what the company looks like. Analysis and insight are about slicing the data ever more finely to identify differences. For example, analysis would show that employee retention decreases with tenure. That's interesting, but not revelatory. Of course more people leave over time. The insight comes when you start comparing very similar groups to tease out what might encourage retention in one and not another. For new salespeople, once you hold constant the effects of performance, pay, and employee level, the big factor that causes retention to drop is lack of promotion. In fact, someone who is not promoted after sixteen quarters is all but guaranteed to quit.

Armed with this insight, you can start predicting the future. Now we know that when the timing of someone's promotion lags behind his peers, he's more likely to quit. And not just that, a more sophisticated analysis might reveal how much more likely he is to quit, and that the greatest increase in risk is after seven or eight quarters. So now you can act.

Most companies, including Google until a few years ago, celebrate promotions but do nothing to reach out to the people who just missed the cut. Which is madness. It takes an hour or two to spot the folks you think will be upset and talk to them about how to continue growing. It's the way you would want to be treated. It's more procedurally just, which helps people perceive the process as more open and honest. It's far better for the company than having someone quit, losing their productivity while you look for a replacement, recruit someone, and then bring them up to speed. And, at a very vulnerable time in someone's career, you're helping him understand what happened and using a demotivating event to ignite his drive.

It takes time to build this kind of capability, but it's easy to begin, regardless of the size of your organization. Start small. Hire a PhD or two in organizational psychology, psychology, or sociology fresh out of grad school. Or bring in a finance or operations person and challenge them to prove that your programs make a difference. Just make sure they are great with statistics and curious about people issues.

Be open to crazy ideas. Find some way to say yes. The ultimate source of innovation for us is Googlers across the company. Google-geist generates hundreds of thousands of comments and ideas, and Googlers are not shy about telling us what they think throughout the year. Vital programs such as equalizing benefits for domestic partners, our child-care centers, and our meditation courses all started with Googlers.

Then experiment. One of the virtues of size is that it creates more opportunities for playing around with the information we've gathered. With about fifty thousand employees, we can carve out a pocket of two hundred or even two thousand people to test ideas. When we changed our performance management system, discussed in chapter 7, we identified early-adopter groups of a few hundred for the first test and over five thousand for the second test before rolling out the changes to the entire company. But even an experiment run on five or ten people is better than nothing. Test an idea on just one team. Or try the idea on the whole company at once, but announce that it's a test for just a month and then you'll decide whether to make it permanent based on how people react. Whether or not Donna Morris and Adobe's experiment of throwing out their performance management system will work or not, I applaud their taking a chance.

Improve relentlessly

In each of the past five years we've improved productivity, as measured by the number of People Operations staff supporting every thousand Googlers, by 6 percent. That may not seem like much, but it means that we are now delivering more services at higher quality with 73 percent of the cost structure we had five years ago on a per-Googler basis. (We're investing more money overall, but far less than we were once you adjust for Google's bigger size today.) Virtually everything we do more than once is measured and improved over time.

We've accomplished this without outsourcing or increasing the use of consultants or vendors. In fact, we've brought more services in-house, which has two virtues. First, it's often cheaper, especially for areas like recruiting and training. Second, there's tremendously useful information to be gleaned by managing processes in-house. In recruiting, for example, we have a central system where recruiters log each candidate interaction, making it easier to reconnect with people who might have turned us down in the past. We've also been able to notice patterns that reveal candidate fraud, such as when one person submitted different resumes under three names, using variations of his first and middle names in the hope of increasing his chances of getting an interview.

Like "HR that just works," running your HR department or team with the same standards of clear objectives, continuous improvement, and reliability that are used in the rest of your company will make your organization credible and trusted.

Field an unconventional team

Let's face it, the HR profession is not held in the highest regard. In 2012, I co-taught a brief course at Stanford University's Graduate School of Business with Professor Frank Flynn, and met an MBA student who aspired to go into HR after business school because she "liked people." As she and I spoke, we realized that she was the only MBA student out of hundreds in her year with that goal. I half-joked that she was probably admitted just to make the class more diverse. There are certainly examples of strong HR leaders and teams, but it's not usually where the cool kids are. When we're children, we want to be firemen or doctors or astronauts. No one wants to be in personnel.

My diagnosis, in part, is that the profession doesn't have the right mix of talent in it, which creates a vicious cycle where the most talented people, who want to work with other talented people, shy away from the field. In too many companies, HR is where you park the nice people who aren't delivering elsewhere. And while most HR professionals are thoughtful, hardworking people, we can all think of colleagues whose greatest strength seemed to be coasting along, barely delivering, just under the radar of management. In 2004 (and I time-stamp this only to prove this was not in the days of slide rules and adding machines), I had such a colleague at GE. She was preparing a spreadsheet to send to her boss, and I was advising her to adjust someone's salary to $106,000 from $100,000 per year. She typed "100" into the spreadsheet, and then "106" in the cell below it. She then took out her calculator, and entered 106 divided by 100, looked at the result, and manually typed "6 percent" into her spreadsheet. She had no idea that one of a spreadsheet's functions is to do the calculating for you. We need to watch for, and take action on, the two tails in our own profession too.

Her story explains why more and more companies are pulling non-HR people into head-of-HR roles. Jodee Kozlak, CHRO of Target, is an attorney by training, as was Allen Hill, who recently retired from the same role at UPS. (Both are friends of mine, and both are brilliant at their jobs.) Microsoft's head of HR, Lisa Brummel, grew up in product management, eBay's Beth Axelrod was a consultant, and Palantir's Michael Lopp was an engineer. CEOs want a business orientation and analytical skill set that is harder to find in HR than it ought to be.

At Google we've built a different kind of People Operations team by applying an unconventional "three-thirds" hiring model. No more than one-third of our hires in People Operations come from traditional HR backgrounds. The core HR expertise they bring is irreplaceable. In addition, they excel at recognizing patterns (such as sensing the difference between a team that is unhappy because a new manager is appropriately turning around a group of underperformers and a team that is unhappy because a new manager is a jerk), build strong relationships at all levels of an organization, and have superb emotional intelligence.

One-third of our hires are recruited from consulting, and specifically from top-tier strategy consultancies, not HR consultancies. I prefer strategy consultants because they have a deep understanding of business and excel at figuring out how to approach and then solve difficult problems. We draw HR expertise from people who have worked as practitioners, so we don't need a double dose of it from the consultants. The consultants also tend to be strong communicators, but we have to screen heavily for emotional intelligence. Having been a consultant, I can attest that consulting firms hire for IQ first and EQ second.[lxx] That's reasonable for them, but in People Operations we need people who can solve problems and also develop deep rapport with a wide range of people across the business. People with higher emotional intelligence also tend to be more self-aware, and hence less arrogant. That also makes it easier for them to move into a new field.

The final third of hires are deeply analytic, holding at least a master's degree in analytical fields ranging from organizational psychology to physics. They keep us honest. They hold our work to a high research standard, and teach the entire team techniques that would otherwise be out of reach for a traditional HR team, such as using the programming languages SQL or R, or methods for coding qualitative data gathered from employee interviews.

The consultants and analysts are also a tremendous source of industry knowledge, familiar with a wide range of other companies and academia, which gives us a jumping-off point for our own work. In a sense, we don't need to hire consulting firms because we've already built one in-house.

[lxx] EQ, or "emotional quotient," is shorthand for the phrase "emotional intelligence."

And then, of course, we mix the groups. Regardless of background, everyone has the opportunity to work in every job, making their days more stimulating, their careers more fulfilling, our team stronger, and our products better. Ex-consultant Judy Gilbert has run recruiting and learning, and is currently the People Operations leader for YouTube and Google[x]. Janet Cho, who worked in finance before moving into HR, has run our mergers-and-acquisitions teams and leads People Operations for our technical organizations. Nancy Lee was a lawyer whose first People Operations job was leading the team that supported Product Management, partnering with Susan Wojcicki, Salar Kamangar, Marissa Mayer, and Jonathan Rosenberg, and who now leads our diversity and education efforts.

By applying our three-thirds model, we recruit a portfolio of capabilities: HR people teach us about influencing and recognizing patterns in people and organizations; the consultants improve our understanding of the business and the level of our problem solving; the analytics people raise the quality of everything we do.

There is little in this book that we could have accomplished without this combination of talent. In the HR profession, it is an error to hire only HR people.

Yet everyone in People Operations has a few traits in common. Each one is a gifted problem solver. Each has a dose of intellectual humility, which makes them open to the possibilities that they could be wrong and always have more to learn. And each is tremendously conscientious, caring deeply about Googlers and the company.

They are a diverse group. The People Operations team collectively speaks more than thirty-five languages and includes former professional athletes, Olympians, world record holders, and veterans. Every major nationality, religion, sexual orientation, and level of physical ability is represented. There are people who have founded their own companies, worked at nonprofits like Teach for America and Catalyst, or joined from other technology companies and industries, and some who have only ever worked at Google. Before joining People Operations, some were engineers, some salespeople, some came from finance and some from public relations or our legal department. Some even worked in human resources! We have people with multiple PhDs and people without any college degrees, and dozens of people who are the first in their families to attend college. It is a phenomenal group of people and it's humbling and a privilege to get to work with them.

But we started as just a handful of people. By diligently applying a high-quality bar and hewing to our three-thirds model, we've built something special over the last nine years or so. You can too. It starts with a clear-eyed assessment of your current portfolio of skills and identifying where you're strong and where you can build. And then it's up to you whom you hire next.

People Operations vs. HR

Shona's branding instincts were superb. Since we created the People Operations name, it's become a popular name for HR departments: Dropbox, Facebook, LinkedIn, Square, Zynga, and over twenty other companies have adopted the moniker.

I recently met the head of People Operations for another technology company. I asked him what made them use that name. He told me, "Oh, it's just regular HR. We just like calling it that."

A part of my heart died in that moment.

Of course, people can use whatever names they want. But they're missing an opportunity to build something different, something perhaps better.

More than anything, what unites us in People Operations is a vision that work doesn't need to be miserable. That it can be ennobling and energizing and exciting. This is what drives us.

That's not to say we have all the answers. We don't. In fact, we have far more questions than answers. But we aspire to bring more insight, innovation, and anticipation to Googlers and how they experience work. It's been dizzying and humbling to be recognized in so many countries and by so many communities as the best place to work. It's been a delight to see people move on from Google and build on what they've learned by creating their own versions of amazing workplaces, including Randy Knaflic (the CHRO of Jawbone), Michael DeAngelo (head of people for Pinterest), Renee Atwood (head of people at Uber), Arnnon Geshuri (the head of HR for Tesla), and Caroline Horn, a partner at venture capital firm Andreessen Horowitz.

I once had a Googler ask me, "If we tell everyone all our people secrets, won't they copy us? Won't we lose our advantage?"

I told him it wouldn't hurt us. "Getting better at recruiting, for example, doesn't mean that you'll hire more people. It means that you'll get better at identifying which people will be more successful in your company. We want the people who will perform their best here, not the ones who will perform better elsewhere."

And if, along the way, work in some companies stops being a means to an end and instead becomes a source of fulfillment, of happiness? If at the end of the day people feel energized and proud of what they've accomplished?

That's okay too.

Work Rules

Chapter 1

WORK RULES...FOR BECOMING A FOUNDER
- ☐ Choose to think of yourself as a founder.
- ☐ Now act like one.

Chapter 2

WORK RULES...FOR BUILDING A GREAT CULTURE
- ☐ Think of your work as a calling, with a mission that matters.
- ☐ Give people slightly more trust, freedom, and authority than you are comfortable giving them. If you're not nervous, you haven't given them enough.

Chapter 3

WORK RULES...FOR HIRING (THE SHORT VERSION)
- ☐ Given limited resources, invest your HR dollars first in recruiting.
- ☐ Hire only the best by taking your time, hiring only people who are better than you in some meaningful way, and not letting managers make hiring decisions for their own teams.

Chapter 4

WORK RULES...FOR FINDING EXCEPTIONAL CANDIDATES
- ☐ Get the best referrals by being excruciatingly specific in describing what you're looking for.

☐ Make recruiting part of everyone's job.

☐ Don't be afraid to try crazy things to get the attention of the best people.

Chapter 5

WORK RULES...FOR SELECTING NEW EMPLOYEES

☐ Set a high bar for quality.

☐ Find your own candidates.

☐ Assess candidates objectively.

☐ Give candidates a reason to join.

Chapter 6

WORK RULES...FOR MASS EMPOWERMENT

☐ Eliminate status symbols.

☐ Make decisions based on data, not based on managers' opinions.

☐ Find ways for people to shape their work and the company.

Chapter 7

WORK RULES...FOR PERFORMANCE MANAGEMENT

☐ Set goals correctly.

☐ Gather peer feedback.

☐ Use a calibration process to finalize ratings.

☐ Split rewards conversations from development conversations.

Chapter 8

WORK RULES...FOR MANAGING YOUR TWO TAILS

☐ Help those in need.

☐ Put your best people under a microscope.

☐ Use surveys and checklists to find the truth and nudge people to improve.

☐ Set a personal example by sharing and acting on your own feedback.

Chapter 9

WORK RULES...FOR BUILDING A LEARNING INSTITUTION

☐ Engage in deliberate practice: Break lessons down into small, digestible pieces with clear feedback and do them again and again.

☐ Have your best people teach.

☐ Invest only in courses that you can prove change people's behavior.

Chapter 10

WORK RULES...FOR PAYING UNFAIRLY

☐ Swallow hard and pay unfairly. Have wide variations in pay that reflect the power law distribution of performance.

☐ Celebrate accomplishment, not compensation.

☐ Make it easy to spread the love.

☐ Reward thoughtful failure.

Chapter 11

WORK RULES...FOR EFFICIENCY, COMMUNITY, AND INNOVATION

☐ Make life easier for employees.

☐ Find ways to say yes.

☐ The bad stuff in life happens rarely...be there for your people when it does.

Chapter 12

WORK RULES...FOR NUDGING TOWARD HEALTH, WEALTH, AND HAPPINESS

☐ Recognize the difference between what is and what ought to be.

☐ Run lots of small experiments.

☐ Nudge, don't shove.

Chapter 13

WORK RULES...FOR SCREWING UP

☐ Admit your mistake. Be transparent about it.

☐ Take counsel from all directions.

☐ Fix whatever broke.

☐ Find the moral in the mistake, and teach it.

Chapter 14

WORK RULES

1. Give your work meaning.
2. Trust your people.
3. Hire only people who are better than you.
4. Don't confuse development with managing performance.
5. Focus on the two tails.
6. Be frugal and generous.
7. Pay unfairly.
8. Nudge.
9. Manage the rising expectations.
10. Enjoy! And then go back to No. 1 and start again.

Acknowledgments

This book quite literally would not have been possible without the unparalleled vision, ambition, and support of Larry Page and Sergey Brin. It's a privilege to learn from them and partner with them. And I'm grateful for their willingness to allow some of Google's lessons to be shared with the world. Eric Schmidt taught me valuable lessons at every one of his staff meetings. A five-minute hallway conversation with him is a master class in leadership. Jonathan Rosenberg, David Drummond, and Shona Brown helped me come up to speed at Google, holding me and the team to ever-higher standards that seemed unachievable at the time but in retrospect were exactly what Google needed. Alan Eustace, Bill Coughran, Jeff Huber, and Urs Hölzle have always been generous with their time and insightful in their arguments. Patrick Pichette has been a great partner in brainstorming as well as a handy source of transportation. I'm indebted to Susan Wojcicki, Salar Kamangar, Stacy Sullivan, Marissa Mayer, and Omid Kordestani for building this place from the ground up and fighting for our culture. And I would have been lost over the years without the sage advice of coach Bill Campbell and Kent Walker's wise counsel.

Three Googlers were crazy enough to think a project like this was worth spending their personal time on it. This book wouldn't be in your hands without Annie Robinson's keen ear for language and research, Kathryn Dekas's analytic brilliance, and Jen Lin's eye for design and clarity. I could write another book just thanking Hannah Cha for all she does to support me and People Operations. My work and my life would be a disaster without you. And thanks to Anna Fraser, Tessa Pompa, Craig Rubens, Prasad Setty, Sunil Chandra, Becky Bucich, Carrie Farrell, Marc Ellenbogen, Scott Rubin, Amy Lambert, Andy Hinton, Kyle Keogh, Rachel Whetstone, Marvin Chow, Miles Johnson, Mimi Kravetz, Leslie Hernandez, Chris Iannuccilli and Lorraine Twohill for their support and advice.

Author Ken Dychtwald gave me the courage to try to track down Amanda Urban, the best literary agent on the planet. Binky, your advocacy, ideas, and courage are unmatched. Thank you. (And my thanks to Ken Auletta for being kind enough to introduce us!)

Courtney Hodell, who fires off emails that are better written than pages on which I spend hours, has been an invaluable editor, ever ready with words of encouragement. If you enjoyed reading this book, thank her. If you didn't, blame me for not listening to her!

I'm grateful to Sean Desmond and Deb Futter, along with Libby Burton, at Twelve for rolling the dice on this first-time author. Hope it works out! :) The lucid prose of *The New Yorker* was my warm-up each morning before writing. It's a superlative magazine that everyone should read...in print!

To the handful of friends who I tortured with early manuscripts, thank you. Craig Bida, Joel Aufrecht, and Adam Grant probably wrote more words of feedback than I wrote for the book, and Cade Massey and Amy Wrzesniewski provided great suggestions.

Gus Mattammal, thanks for helping me knit a coherent thesis from a tangle of ideas, and for helping me brainstorm and refine every MWF. Jason Corley, I never thought those years of writing 1ACs together would turn out to be useful! Between you two, John Busenberg, and Craig I couldn't ask for better friends.

To my parents, Susan Bonczos and Paul Bock, you literally risked everything to give us freedom. All I've done flows from that moment of courage in Romania and your untiring work and support in all the years since. Steve, thanks for being at my side from the beginning. I always knew you would be there if I needed you. I love you all.

In life you have two families: the one you're born into and the one you choose. I'm the luckiest guy on the planet because Gerri Ann chose to make me part of hers. To paraphrase, I realized, ever since I met you, every single day of my life has been better than the day before it. So that means every single day that you see me, that's on the best day of my life. You and the girls let me steal nights and weekends from you to write this book. I love you all more than anything. And I can't wait for the next weekend!

Finally, thank you to the remarkable Googlers I get to work with each day and to the amazing, amazing People Operations team. I've told them before and I'll keep saying it: It's a privilege to be able to work alongside you, learn from you, create with you. There's not a team like you on the planet, and it's a gift to enjoy your company.

Photo Credits

Page 30: Google Images & Tessa Pompa

Page 31: Google & Burning Man

Page 32: The Google Doodle team

Page 35: Google Maps

Page 35: Google Maps

Page 36: Google Maps

Page 36: Google Maps (maps.google.com/oceans)

Page 37: Google Maps

Page 38: Google Maps

Page 43: Change.gov

Page 73: Google

Page 73: Google

Page 74: Google

Page 95: Google

Page 104: Google

Page 105: Google

Page 109: Google

Page 111: Google

Page 116: Google Creative Lab

Page 127: Photo courtesy of Brett Crosby

Page 129: Inspired by Adam Wald

Page 139: Google

Page 140: Google

Page 158: Paul Cowan

Page 162: Google

Page 166: Google

Page 172: Google

Page 173: Google

Page 179: Courtesy of Archives & Special Collections at the Thomas J. Dodd Research Center, University of Connecticut Libraries

Page 179: Courtesy of Archives & Special Collections at the Thomas J. Dodd Research Center, University of Connecticut Libraries

Page 179: Tessa Pompa

Page 181: Google

Page 191: Google

Page 192: Google

Page 195: Google

Page 197: Google

Page 198: Google

Page 211: Google

Page 227: Google

Page 227: Google

Page 228: Lycos

Page 228: Mindspark/Excite

Page 229: Google

Page 246: Tessa Pompa & Diana Funk

Page 250: Google

Page 251: Craig Rubens & Tessa Pompa

Page 254: Google

Page 270: Google

Page 274: Google

Page 283: Photo Sphere image courtesy of Noam Ben-Haim

Page 287: Courtesy of rAndom International

Page 289: Photo by Hiroko Masuike, The New York Times, 3/22/13

Page 300: Google

Page 302: Courtesy of Manu Cornet

Page 304: Hachette/Publisher

Page 310: Courtesy of Prof. David Hammond, PhD, University of Waterloo

Page 311: Google

Page 314: Inspired by the Delboeuf illusion

Page 351: Google

Page 385: Tessa Pompa

Notes

1. US Bureau of Labor Statistics, "Charts from the American Time Use Survey," last modified October 23, 2013, http://www.bls.gov/tus/charts/.

2. John A. Byrne, "How Jack Welch Runs GE," *BusinessWeek*, June 8, 1998, http://www.businessweek.com/1998/23/b3581001.htm.

3. (Name protected), confidential communications to author, 2006–2007.

4. Will Oremus, a senior technology writer at Slate, had this to add about how revolutionary Gmail was: "Ten years on, Google's rivals have copied Gmail so thoroughly that it's hard to remember just how terrible webmail was before Gmail came along. Pages were clunky and slow to load, search functions were terrible, and spam was rampant. You couldn't organize messages by conversation. Storage capacity was anemic, and if you ran out of space, you had to spend hours deleting old emails or buy more storage from your provider. Gmail, which was designed using Ajax rather than plain old HTML, taught us that Web apps could run as smoothly as desktop applications. And it taught us the power of cloud storage." April 1, 2014, http://www.slate.com/blogs/future_tense/2014/04/01/gmail_s_10th_birthday_the_google_april_fool_s_joke_that_changed_tech_history.html.

5. James Raybould, "Unveiling LinkedIn's 100 Most InDemand Employers of 2013," *LinkedIn* (official blog), October 16, 2013, http://blog.linkedin.com/2013/10/16/unveiling-linkedins-100-most-indemand-employers-of-2013/.

6. Our actual hiring numbers vary from year to year.

7. "Harvard Admitted Students Profile," Harvard University, accessed January 23, 2014, https://college.harvard.edu/admissions/admissions-statistics.

8. "Yale Facts and Statistics," Yale University, accessed January 23, 2014, http://oir.yale.edu/sites/default/files/FACTSHEET(2012-13)_3.pdf.

9. "Admission Statistics," Princeton University Undergraduate Admission, accessed January 23, 2014, http://www.princeton.edu/admission/applyingforadmission/admission_statistics/.

10. Source: Google, Inc.

11. "Fortune's 100 Best Companies to Work For®," Great Place to Work Institute, accessed January 23, 2014, http://www.greatplacetowork.net/best-companies/north-america/united-states/fortunes-100-best-companies-to-work-forr/441-2005.

12. "Wegmans Announces Record Number of Employee Scholarship Recipients in 2012," Wegmans, June 7, 2012, https://www.wegmans.com/webapp/wcs/stores/servlet/PressReleaseDetailView?productId=742304&storeId=10052&catalogId=10002&langId=-1.

13. Sarah Butler and Saad Hammadi, "Rana Plaza factory disaster: Victims still waiting for compensation," theguardian.com, October 23, 2013, http://www .theguardian.com/world/2013/oct/23/rana-plaza-factory-disaster-compensation -bangladesh.

14. *Office Space*, directed by Mike Judge (1999; 20th Century Fox).

15. Richard Locke, Thomas Kochan, Monica Romis, and Fei Qin, "Beyond Corporate Codes of Conduct: Work Organization and Labour Standards at Nike's Suppliers," *International Labour Review* 146, no. 1–2 (2007): 21–40.

16. Kamal Birdi, Chris Clegg, Malcolm Patterson, Andrew Robinson, Chris B. Stride, Toby D. Wall, and Stephen J. Wood, "The Impact of Human Resource and Operational Management Practices on Company Productivity: A Longitudinal Study," *Personnel Psychology* 61 (2008): 467–501.

17. The four were Francis Upton, Charles Batcheldor, Ludwig Boehm, and John Kruesi. See "Six teams that changed the world," *Fortune*, May 31, 2006, http://money.cnn.com/2006/05/31/magazines/fortune/sixteams_greatteams_for tune_061206/.

18. Nicole Mowbray, "Oprah's path to power," *The Observer*, March 2, 2003, http://www.theguardian.com/media/2003/mar/02/pressandpublishing.usnews1.

19. Adam Lashinsky, "Larry Page: Google should be like a family," *Fortune*, January 19, 2012, http://fortune.com/2012/01/19/larry-page-google-should-be-like -a-family/.

20. Larry Page's University of Michigan Commencement Address, http:// googlepress.blogspot.com/2009/05/larry-pages-university-of-michigan.html.

21. Mark Malseed, "The Story of Sergey Brin," *Moment*, February–March 2007, http://www.momentmag.com/the-story-of-sergey-brin/.

22. Steven Levy, *In the Plex: How Google Thinks, Works, and Shapes Our Lives* (New York: Simon & Schuster, 2011).

23. John Battelle, "The Birth of Google," *Wired*, August 2005, http://www .wired.com/wired/archive/13.08/battelle.html. "Our History in Depth," Google, http://www.google.com/about/company/history/.

24. Those investments would eventually be worth in excess of $1 billion. Each. Bechtolsheim's and Cheriton's investments weren't the original source of support for what would eventually become Google. The National Science Foundation (NSF) was an even earlier underwriter, albeit less directly, through its Digital Libraries Initiative (DLI). Professors Hector Garcia-Molina and Terry Winograd received a DLI grant on September 1, 1994, to create the Stanford Integrated Digital Library Project. It's mission was "to develop the enabling technologies for a single, integrated and 'universal' library, proving uniform access to the large number of emerging networked information sources and collections … [including] everything from personal information collections, to the collections that one finds today in conventional libraries, to the large data collections shared by scientists. The technology developed in this project will provide the 'glue' that will make this worldwide collection usable as a unified entity, in a scalable and economically viable fashion."

Larry's graduate studies were supported by this grant, as was some of the early work that would become Google. Sergey's studies were also supported by an NSF Graduate Research Fellowship. See "On the Origins of Google," National Science Foundation, http://www.nsf.gov/discoveries/disc_summ.jsp?cntn_id=100660.

25. "Code of Conduct," Google, http://investor.google.com/corporate/code -of-conduct.html#II.

26. Henry Ford, *My Life and Work* (Garden City, NY: Doubleday, Page, 1922).

27. Hardy Green, *The Company Town: The Industrial Edens and Satanic Mills That Shaped the American Economy* (New York: Basic Books, 2010).

28. "About Hershey: Our Proud History," Hershey Entertainment & Resorts, http://www.hersheypa.com/about_hershey/our_proud_history/about_milton _hershey.php.

29. *American Experience*: "Henry Ford," WGBH Educational Foundation, first broadcast March 2013. Written, produced, and directed by Sarah Colt. See also: Albert Lee, *Henry Ford and the Jews* (New York: Stein and Day, 1980). Ford also owned the *Dearborn Independent*, a weekly newspaper that ran regular anti-Semitic articles and editorials, some under Ford's byline, a number of which were published in four volumes, collectively known as *The International Jew: the World's Problem* (1920–1922).

30. Michael D. Antonio, *Hershey: Milton S. Hershey's Extraordinary Life of Wealth, Empire, and Utopian Dreams* (New York: Simon & Schuster, 2007). The Hershey town newspaper, first printed when the town had only 250 people, included occasional advice such as "Race suicide [i.e., racial intermarriage] is the greatest evil of the day," from a column describing "Mrs. George Herrick, mother of New England's first 'eugenic' baby" (*Hershey's Weekly*, December 26, 1912). Similarly, the Milton Hershey School was established for "poor, healthy, white, male orphans" (Milton Hershey School Deed of Trust, November 15, 1909).

31. Jon Gertner, "True Innovation," *New York Times*, February 25, 2012, http://www.nytimes.com/2012/02/26/opinion/sunday/innovation-and-the-bell -labs-miracle.html?pagewanted=all&_r=0.

32. Jon Gertner, *The Idea Factory: Bell Labs and the Great Age of American Innovation*, reprint edition (New York: Penguin, 2013).

33. John R. Pierce, "Mervin Joe Kelly, 1894–1971" (Washington, DC: National Academy of Sciences, 1975), http://www.nasonline.org/publications /biographical-memoirs/memoir-pdfs/kelly-mervin.pdf.

34. "Google Search Now Supports Cherokee," *Google* (official blog), March 25, 2011, http://googleblog.blogspot.com/2011/03/google-search-now-supports -cherokee.html.

35. "Some Weekend Work That Will (Hopefully) Enable More Egyptians to Be Heard," *Google* (official blog), January 31, 2011, http://googleblog.blogspot .com/2011/01/some-weekend-work-that-will-hopefully.html.

36. Lashinsky, "Larry Page: Google should be like a family."

37. Edgar H. Schein, *Organizational Culture and Leadership* (San Francisco: Jossey-Bass, 2010).

38. Googlegeist (our annual employee survey), 2013.

39. The *RescueTime* blog estimated that visitors to our website that day spent a total of 5,350,000 hours (twenty-six seconds each) playing Les Paul's guitar. "Google Doodle Strikes Again! 5.35 Million Hours Strummed," *RescueTime* (blog), June 9, 2011, http://blog.rescuetime.com/2011/06/09/google-doodle-strikes-again/.

40. Compare this to the mission of the Stanford Integrated Digital Library from note 24. Sounds familiar, right?

41. "IBM Mission Statement," http://www.slideshare.net/waqarasif67/ibm-mission-statement.

42. "Mission & Values," McDonald's, http://www.aboutmcdonalds.com/mcd/our_company/mission_and_values.html.

43. "The Power of Purpose," Proctor & Gamble, http://www.pg.com/en_US/company/purpose_people/index.shtml.

44. *Wikipedia*, "Timeline of Google Street View," last modified May 19, 2014, http://en.wikipedia.org/wiki/Timeline_of_Google_Street_View.

45. South Base Camp, Mt. Everest, https://www.google.com/maps/@28.00 7168,86.86105,3a,75y,92.93h,87.22t/data=!3m5!1e1!3m3!1sUdU6omw_CrN8sm7N WUnpcw!2eo!3e2.

46. Heron Island, https://www.google.com/maps/views/streetview/oceans ?gl=us.

47. Philip Salesses, Katja Schechtner, and César A. Hidalgo, "The Collaborative Image of the City: Mapping the Inequality of Urban Perception," *PLOS ONE*, July 24, 2013, http://www.plosone.org/article/info%3Adoi%2F10.1371%2 Fjournal.pone.0068400.

48. Full disclosure: Google Capital has invested in Uber. And since I started on this manuscript, Google has acquired Waze.

49. "Google Developers," Google, May 15, 2013, https://plus.sandbox. google.com/+GoogleDevelopers/posts/NrPMMwZtY8m.

50. Adam Grant, *Give and Take: A Revolutionary Approach to Success* (New York: Viking, 2013).

51. Adam M. Grant, Elizabeth M. Campbell, Grace Chen, Keenan Cottone, David Lapedis, and Karen Lee, "Impact and the Art of Motivation Maintenance: The Effects of Contact with Beneficiaries on Persistence Behavior," *Organizational Behavior and Human Decision Processes* 103, no. 1 (2007): 53–67.

52. Corey Kilgannon, "The Lox Sherpa of Russ & Daughters," *New York Times*, November 2, 2012, http://www.nytimes.com/2012/11/04/nyregion/the-lox -sherpa-of-russ-daughters.html?_r=0, 11/2/12.

53. A. Wrzesniewski, C. McCauley, R. Rozin, and B. Schwarz, "Jobs, Careers, and Callings: People's Relation to Their Work," *Journal of Research in Personality* 31 (1997): 21–33.

54. Roger More, "How General Motors Lost its Focus—and its Way," *Ivey Business Journal*, May–June 2009, http://iveybusinessjournal.com/topics/strategy /how-general-motors-lost-its-focus-and-its-way#.UobINPlwp8E.

55. Marty Makary, MD, *Unaccountable: What Hospitals Won't Tell You and How Transparency Can Revolutionize Health Care* (New York: Bloomsbury Press, 2012).

56. Daniel Gross, "Bridgewater May Be the Hottest Hedge Fund for Harvard Grads, but It's Also the Weirdest," *Daily Beast*, March 7, 2013, http://www.thedailybeast.com/articles/2013/03/07/bridgewater-may-be-the-hottest-hedge-fund-for-harvard-grads-but-it-s-also-the-weirdest.html.

57. "Radical Transparency," *LEADERS*, July–September 2010, http://www.leadersmag.com/issues/2010.3_Jul/Shaping%20The%20Future/Ray-Dalio-Bridgewater-Associates-Interview-Principles.html.

58. Tara Siegel Bernard, "Google to Add Pay to Cover a Tax for Same-Sex Benefits," *New York Times*, June 30, 2010, http://www.nytimes.com/2010/07/01/your-money/01benefits.html?_r=2&.

59. Ethan R. Burris, "The Risks and Rewards of Speaking Up: Managerial Responses to Employee Voice," *Academy of Management Journal* 55, no. 4 (2012): 851–875. Robert S. Dooley and Gerald E. Fryxell, "Attaining Decision Quality and Commitment from Dissent: The Moderating Effects of Loyalty and Competence in Strategic Decision-Making Teams," *Academy of Management Journal* 42, no. 4 (1999): 389–402. Charlan Jeanne Nemeth, "Managing Innovation: When Less Is More," *California Management Review* 40, no. 1 (1997): 59–74. Linda Argote and Paul Ingram, "Knowledge Transfer: A Basis for Competitive Advantage in Firms," *Organizational Behavior and Human Decision Processes* 82, no. 1 (2000): 150–169.

60. "China Blocking Google," BBC News World Edition, September 2, 2002, http://news.bbc.co.uk/2/hi/technology/2231101.stm.

61. Laszlo Bock, "Passion, Not Perks," Google, September 2011, http://www.thinkwithgoogle.com/articles/passion-not-perks.html.

62. According to Bridget Lawlor, the archivist for the Drucker Institute, "[This] quote…is often attributed to Drucker but we do not have a definitive source. It is certainly possible that he used it in a speech or lecture but we do not have a transcript of it."

63. Stacy was on to something. Simon Lam of the University of Hong Kong and John Schaubroeck of Michigan State University found that selecting frontline "opinion leaders" to implement changes and guide norms has a greater impact than using managers or random employees. They looked at three bank branches rolling out a new service training program. When "opinion leaders" were selected as "service quality leaders," customers, supervisors, and tellers all observed significantly greater improvements in service quality, even though the quality training was otherwise identical. Simon S. Lam and John Schaubroeck, "A Field Experiment Testing Frontline Opinion Leaders as Change Agents," *Journal of Applied Psychology* 85, no. 6 (2000): 987–995.

64. Pre-2013 salary data from http://www.stevetheump.com/Payrolls.htm and http://www.baseballprospectus.com/compensation/cots/. Salary data for 2013 from *USA Today*, accessed December 15, 2013, http://www.usatoday.com/sports/fantasy/baseball/salaries/2013/all/team/all/. Winner data from http://espn.go.com/mlb/worldseries/history/winners.

65. David Waldstein, "Penny-Pinching in Pinstripes? Yes, the Yanks Are Reining in Pay," *New York Times*, March 11, 2013, http://www.nytimes.com

/2013/03/12/sports/baseball/yankees-baseballs-big-spenders-are-reining-it-in
.html?pagewanted=all&_r=0.

66. "Milestones in Mayer's Tenure as Yahoo's Chief," *New York Times*, January 16, 2014, http://www.nytimes.com/interactive/2014/01/16/technology /marissa-mayer-yahoo-timeline.html?_r=0#/#time303_8405.

67. Brian Stelter, "He Has Millions and a New Job at Yahoo. Soon, He'll Be 18," *New York Times*, March 25, 2013, http://www.nytimes.com/2013/03/26/business /media/nick-daloisio-17-sells-summly-app-to-yahoo.html?hp&_r=0. Kevin Roose, "Yahoo's Summly Acquisition Is About PR and Hiring, Not a 17-Year-Old's App," *New York*, March 26, 2013, http://nymag.com/daily/intelligencer/2013/03/yahoos -summly-acquisition-is-about-image.html.

68. "Yahoo Acquires Xobni App," Zacks Equity Research, July 5, 2013, http://finance.yahoo.com/news/yahoo-acquires-xobni-app-154002114.html.

69. Professor Freek Vermeulen, "Most Acquisitions Fail—Really," *Freeky Business* (blog), January 3, 2008, http://freekvermeulen.blogspot.com/2007/11 /random-rantings-2.html.

70. Ambady and Rosenthal, "Thin Slices," among many others. Detailed citations are in the following chapter and start with endnote 80. Some researchers have found that impressions are formed in as little as ten seconds.

71. Caroline Wyatt, "Bush and Putin: Best of Friends," BBC News, June 16, 2001, http://news.bbc.co.uk/2/hi/1392791.stm.

72. While experts disagree on exactly how much training is wasted, they are near unanimous in agreeing that very little training has an effect. John Newstrom (1985) surveyed members of the American Society of Trainers and Developers, who estimated that about 40 percent of training is applied immediately afterward, but a year later only 15 percent is still being applied. And I'd point out that the trainers themselves are likeliest to have a rosy estimate. When Newstrom and Mary Broad (1992) looked again, they found that about 20 percent of learners applied their training, though Broad (2005) later clarified that most programs are closer to a 10 percent transfer of learning to actual performance. Tim Baldwin and Kevin Ford (1988) conclude as well that "while American industries annually spend up to $100 billion on training and development, not more that 10% of these expenditures actually result in transfer to the job." Scott Tannenbaum and Gary Yukl (1992) were even less sanguine, estimating that 5 percent of learners apply what they learn.

Recent research from Eduardo Salas, Tannenbaum, Kurt Kraiger, and Kimberly Smith-Jentsch (2012) offers some hope. Training can be much more effective, they argue, but requires certain conditions, including a supportive work environment, formal and informal reinforcement of skills learned in training, job autonomy, organizational commitment to quality, and the flexibility for employees to try out and perform the new, learned behaviors in their jobs. In short, exactly the kind of environment we're discussing how to create.

73. "Einstein at the Patent Office," Swiss Federal Institute of Intellectual Property, last modified April 21, 2011, https://www.ige.ch/en/about-us/einstein /einstein-at-the-patent-office.html.

74. Corporate Executive Board, Corporate Leadership Council, HR Budget and Efficiency Benchmarking Database, Arlington VA, 2012.

75. Pui-Wing Tam and Kevin Delaney, "Google's Growth Helps Ignite Silicon Valley Hiring Frenzy," *Wall Street Journal*, November 23, 2005, http://online .wsj.com/article/SB113271436430704916.html, and personal conversations.

76. Malcolm Gladwell, "The Talent Myth: Are Smart People Overrated?," *The New Yorker*, July 22, 2002, http://www.newyorker.com/archive /2002/07/22/020722fa_fact?currentPage=all.

77. "Warning: We Brake for Number Theory," *Google* (official blog), July 12, 2004, http://googleblog.blogspot.com/2004/07/warning-we-brake-for-number -theory.html.

78. "Google Hiring Experience," *Oliver Twist* (blog), last modified January 17, 2006, http://google-hiring-experience.blogspot.com/.

79. "How Tough Is Google's Interview Process," *Jason Salas' WebLog* (blog), September 5, 2005, http://weblogs.asp.net/jasonsalas/archive/2005/09/04/424378.aspx.

80. The earliest research on this is from B. M. Springbett of the University of Manitoba, published in 1958. Though using a very small sample of interviewers, he found that decisions were typically made within the first four minutes of an interview. Subsequent research includes: Nalini Ambady and Robert Rosenthal, "Thin Slices of Expressive Behavior as Predictors of Interpersonal Consequences: A Meta-Analysis," *Psychological Bulletin* 111, no. 2 (1992): 256–274; M. R. Barrick, B. W. Swider, and G. L. Stewart, "Initial Evaluations in the Interview: Relationships with Subsequent Interviewer Evaluations and Employment Offers," *Journal of Applied Psychology* 95, no. 6 (2010): 1163–1172; M. R. Barrick, S. L. Dustin, T. L. Giluk, G. L. Stewart, J. A. Shaffer, and B. W. Swider, "Candidate Characteristics Driving Initial Impressions During Rapport Building: Implications for Employment Interview Validity," *Journal of Occupational and Organizational Psychology* 85, no. 2 (2012): 330–352.

81. J. T. Prickett, N. Gada-Jain, and F. J. Bernieri, "The Importance of First Impressions in a Job Interview," paper presented at the annual meeting of the Midwestern Psychological Association, Chicago, IL, May 2000.

82. *Wikipedia*, "Confirmation bias," http://en.wikipedia.org/wiki/Confirmation _bias#CITEREFPlous1993, citing Scott, Plous, *The Psychology of Judgment and Decision Making*, (New York: McGraw-Hill, 1993), 233.

83. Gladwell, "The New-Boy Network, *The New Yorker*, May 29, 2000: 68–86.

84. N. Munk and S. Oliver, "Think Fast!" *Forbes*, 159, no. 6 (1997): 146–150. K. J. Gilhooly and P. Murphy, "Differentiating Insight from Non-Insight Problems," *Thinking & Reasoning* 11, no. 3 (2005): 279–302.

85. Frank L. Schmidt and John E. Hunter, "The Validity and Utility of Selection Methods in Personnel Psychology: Practical and Theoretical Implications of 85 Years of Research Findings," *Psychological Bulletin* 124, no. 2 (1998): 262–274. The r^2 values presented in this chapter are calculated based on the reported corrected correlation coefficients (r).

86. Phyllis Rosser, *The SAT Gender Gap: Identifying the Causes* (Washington, DC: Center for Women Policy Studies, 1989).

87. Subsequent studies have validated the gender gap on the SAT and demonstrated racial bias as well. See, for example, Christianne Corbett, Catherine Hill, and Andresse St. Rose, "Where the Girls Are: The Facts About Gender Equity in Education," American Association of University Women (2008). See also Maria Veronica Santelices and Mark Wilson, "Unfair Treatment? The Case of Freedle, the SAT, and the Standardization Approach to Differential Item Functioning, *Harvard Educational Review* 80, no. 1 (2010): 106–134.

88. Alec Long, "Survey Affirms Pitzer Policy Not to Require Standardized Tests," *The Student Life*, February 28, 2014.

89. Michael A. McDaniel, Deborah L. Whetzel, Frank L. Schmidt, and Steven D. Maurer, "The Validity of Employment Interviews: A Comprehensive Review and Meta-Analysis," *Journal of Applied Psychology* 79, no. 4 (1994): 599–616. Willi H. Wiesner and Steven F. Cronshaw, "A Meta-Analytic Investigation of the Impact of Interview Format and Degree of Structure on the Validity of the Employment Interview," *Journal of Occupational Psychology* 61, no. 4 (1988): 275–290.

90. Like any good thing, conscientiousness carried to an extreme can become a negative as it slips from careful planning, goal setting, and persistence into inflexibility and compulsive perfectionism. Thus far we haven't seen this as an issue, but plan to explore it in the future.

91. Personal correspondence, October 7, 2014.

92. Abraham H. Maslow, *The Psychology of Science: A Reconnaissance* (New York: Joanna Cotler Books, 1966), 15.

93. Each candidate received a score from each interviewer on a scale of 0.0 to 4.0, which was then averaged across interviewers to yield an overall score. A 3.0 nominally meant we should hire that person, but as a practical matter almost all candidates who were hired ended up between 3.2 and 3.6. No one has ever averaged a 4.0.

94. David Smith, "Desmond Tutu Attacks South African Government over Dalai Lama Visit," *Guardian*, October 4, 2011, http://www.theguardian.com /world/2011/oct/04/tutu-attacks-anc-dalai-lama-visa.

95. The moving video is here: http://www.youtube.com/watch?v=97bZu -tXLq4.

96. John Emerich Edward Dalberg, Lord Acton, Letter to Bishop Mandell Creighton, April 5, 1887, in *Historical Essays and Studies*, eds. John Neville Figgis and Reginald Vere Laurence (London: Macmillan, 1907), 504.

97. Discovering Psychology with Philip Zimbardo, PhD, updated edition, "Power of the Situation," reference starts at 10 minutes 59 seconds into video, http://www.learner.org/series/discoveringpsychology/19/e19expand.html.

98. There has been extensive research exploring, expanding on, and criticizing Milgram's findings. For example, see works by Alex Haslam (University of Queensland) and Stephen Reicher (University of St. Andrews).

99. Richard Norton Smith, "Ron Nessen," Gerald R. Ford Oral History Project, http://geraldrfordfoundation.org/centennial/oralhistory/ron-nessen/.

100. "SciTech Tuesday: Abraham Wald, Seeing the Unseen," post by Annie Tete, STEM Education Coordinator at the National World War II Museum, *See & Hear* (museum blog), November 13, 2012, http://www.nww2m.com/2012/11 /scitech-tuesday-abraham-wald-seeing-the-unseen/. A reprint of Wald's work can be found here: http://cna.org/sites/default/files/research/0204320000.pdf.

101. "Lawyercat" is Googler-speak for the hardworking and vigilant Googlers in our legal function. And, yes, Googlers often use an actual photo of a cat (complete with suit, tie, and stiff white collar) to accompany internal discussions that might veer into murky legal territory.

102. "Our New Search Index: Caffeine," *Google* (official blog), June 8, 2010, http://googleblog.blogspot.com/2010/06/our-new-search-index-caffeine.html.

103. "Time to Think," 3M, http://solutions.3m.com/innovation/en_US /stories/time-to-think.

104. Ryan Tate, "Google Couldn't Kill 20 Percent Time Even If It Wanted To," *Wired*, August 21, 2013, http://www.wired.com/business/2013/08/20-percent -time-will-never-die/.

105. Linda Babcock, Sara Laschever, Michele Gelfand, and Deborah Small, "Nice Girls Don't Ask," *Harvard Business Review*, October 2003, http://hbr .org/2003/10/nice-girls-dont-ask/. Linda Babcock and Sara Laschever, *Women Don't Ask: Negotiation and the Gender Divide* (Princeton, NJ: Princeton University Press, 2003).

106. "Employee Engagement: What's Your Engagement Ratio?" Gallup Consulting, Employment Engagement Overview Brochure, downloaded 11/17/13.

107. William H. Macey and Benjamin Schneider, "The Meaning of Employee Engagement," *Industrial and Organizational Psychology* 1, no. 1 (2008): 3–30.

108. Olivier Serrat, "The Travails of Micromanagement" (Washington, DC: Asian Development Bank, 2011), http://digitalcommons.ilr.cornell.edu/cgi /viewcontent.cgi?article=1208&context=intl.

109. Richard Bach, *Illusions: The Adventures of a Reluctant Messiah* (New York: Delacorte, 1977).

110. Elaine D. Pulakos and Ryan S. O'Leary, "Why Is Performance Management Broken?" *Industrial and Organizational Psychology* 4, no. 2 (2011): 146–164.

111. "Results of the 2010 Study on the State of Performance Management," Sibson Consulting, 2010, http://www.sibson.com/publications/surveysand studies/2010SPM.pdf.

112. Julie Cook Ramirez, "Rethinking the Review," *Human Resource Executive HREOnline*, July 24, 2013, http://www.hreonline.com/HRE/view/story.jhtml ?id=534355695.

113. Edwin A. Locke and Gary P. Latham, *A Theory of Goal Setting & Task Performance* (Upper Saddle River, NJ: Prentice Hall, 1990).

114. Xander M. Bezuijen, Karen van Dam, Peter T. van den Berg, and Henk Thierry, "How Leaders Stimulate Employee Learning: A Leader-Member Exchange Approach," *Journal of Occupational and Organizational Psychology* 83, no. 3 (2010): 673–693. Benjamin Blatt, Sharon Confessore, Gene Kallenberg, and

Larrie Greenberg, "Verbal Interaction Analysis: Viewing Feedback Through a Different Lens," *Teaching and Learning in Medicine* 20, no. 4 (2008): 329–333.

115. Elaine D. Pulakos and Ryan S. O'Leary, "Why Is Performance Management Broken?" *Industrial and Organizational Psychology* 4, no. 2 (2011): 146–164.

116. For any Googlers reading this, use of this meme was approved both by Paul Cowan and Colin McMillen, as well as GCPA. What happens on Memegen, stays on Memegen!

117. Susan J. Ashford, "Feedback-Seeking in Individual Adaptation: A Resource Perspective," *Academy of Management Journal* 29, no. 3 (1986): 465–487. Leanne E. Atwater, Joan F. Brett, and Atira Cherise Charles, "Multisource Feedback: Lessons Learned and Implications for Practice," *Human Resource Management* 46, no. 2 (2007): 285–307. Roger Azevedo and Robert M. Bernard, "A Meta-Analysis of the Effects of Feedback in Computer-Based Instruction," *Journal of Educational Computing Research* 13, no. 2 (1995): 111–127. Robert A. Baron, "Criticism (Informal Negative Feedback) As a Source of Perceived Unfairness in Organizations: Effects, Mechanisms, and Countermeasures," in *Justice in the Workplace: Approaching Fairness in Human Resource Management* (Applied Psychology Series), ed. Russell Cropanzano (Hillsdale, NJ: Lawrence Erlbaum Associates, Inc., 1993), 155–170. Donald B. Fedor, Walter D. Davis, John M. Maslyn, and Kieran Mathieson, "Performance Improvement Efforts in Response to Negative Feedback: The Roles of Source Power and Recipient Self-Esteem," *Journal of Management* 27, no. 1 (2001): 79–97. Gary E. Bolton, Elena Katok, and Axel Ockenfels, "How Effective Are Electronic Reputation Mechanisms? An Experimental Investigation," *Management Science* 50, no. 11 (2004): 1587–1602. Chrysanthos Dellarocas, "The Digitization of Word of Mouth: Promise and Challenges of Online Feedback Mechanisms," *Management Science* 49, no. 10 (2003): 1407–1424.

118. Edward L. Deci, "Effects of Externally Mediated Rewards on Intrinsic Motivation," *Journal of Personality and Social Psychology* 18, no. 1 (1971): 105–115.

119. Edward L. Deci and Richard M. Ryan, *Intrinsic Motivation and Self-Determination in Human Behavior* (New York: Plenum, 1985). E. L. Deci, R. Koestner, and R. M. Ryan, "A Meta-Analytic Review of Experiments Examining the Effects of Extrinsic Rewards on Intrinsic Motivation," *Psychological Bulletin* 125, no. 6 (1999): 627–668. R. M. Ryan and E. L. Deci, "Self-Determination Theory and the Facilitation of Intrinsic Motivation, Social Development, and Well-Being," *American Psychologist* 55, no. 1 (2000): 68–78.

120. Maura A. Belliveau, "Engendering Inequity? How Social Accounts Create vs. Merely Explain Unfavorable Pay Outcomes for Women," *Organization Science* 23, no. 4 (2012): 1154–1174, published online September 28, 2011, http://pubsonline.informs.org/doi/abs/10.1287/orsc.1110.0691.

121. Personal conversation.

122. Atwater, Brett, and Charles, "Multisource Feedback." Blatt, Confessore, Kallenberg, and Greenberg, "Verbal Interaction Analysis." Joan F. Brett and Leanne E. Atwater, "360° Feedback: Accuracy, Reactions, and Perceptions of Usefulness," *Journal of Applied Psychology* 86, no. 5 (2001): 930–942.

123. The engineers who developed a precursor to this system decided on 512 characters. They originally wanted 256 characters, in part because a byte (which is a collection of binary digits, or "bits") can store one of 256 different values. But then they figured that 256 characters probably wasn't enough so they doubled it. (256 is also 2^8 and 512 is 2^9.)

124. Drew H. Bailey, Andrew Littlefield, and David C. Geary, "The Co-development of Skill at and Preference for Use of Retrieval-Based Processes for Solving Addition Problems: Individual and Sex Differences from First to Sixth Grades," *Journal of Experimental Child Psychology* 113, no. 1 (2012): 78–92.

125. Albert F. Blakeslee, "Corn and Men," *Journal of Heredity* 5, no. 11 (1914): 511–518. See Mark F. Schilling, Ann E. Watkins, and William Watkins, "Is Human Height Bimodal?" *The American Statistician* 56, no. 3 (2002): 223–229, http://faculty.washington.edu/tamre/IsHumanHeightBimodal.pdf.

126.

127. Carl Friedrich Gauss, *Theory of the Motion of the Heavenly Bodies Moving about the Sun in Conic Sections: A Translation of Gauss's "Theoria Motus,"* trans. Charles Henry Davis (1809; repr., Boston: Little, Brown & Co., 1857).

128. Margaret A. McDowell, Cheryl D. Fryar, Cynthia L. Ogden, and Katherine M. Flegal, "Anthropometric Reference Data for Children and Adults: United States, 2003–2006," *National Health Statistics Reports* 10 (Hyattsville, MD: National Center for Health Statistics, 2008), http://www.cdc.gov/nchs/data/nhsr/nhsr010.pdf.

129. Aaron Clauset, Cosma Rohilla Shalizi, and M. E. J. Newman, "Power-Law Distributions in Empirical Data," *SIAM Review* 51, no. 4 (2009): 661–703.

130. Herman Aguinis and Ernest O'Boyle Jr., "Star Performers in Twenty-First Century Organizations," *Personnel Psychology* 67, no. 2 (2014): 313–350.

131. Boris Groysberg, Harvard Business School, http://www.hbs.edu/faculty/Pages/profile.aspx?facId=10650.

132. Note that improving to "average" levels isn't necessarily the same as becoming the fiftieth-best performer (i.e., the median performer), but it's close enough for illustrative purposes.

133. Jack and Suzy Welch, "The Case for 20-70-10," *Bloomberg Businessweek*, October 1, 2006, http://www.businessweek.com/stories/2006-10-01/the-case-for-20-70-10.

134. Ibid.

135. Kurt Eichenwald, "Microsoft's Lost Decade," *Vanity Fair,* August 2012, http://www.vanityfair.com/business/2012/08/microsoft-lost-mojo-steve-ballmer.

136. Tom Warren, "Microsoft Axes Its Controversial Employee-Ranking System," *The Verge,* November 12, 2013, http://www.theverge.com/2013/11/12/5094 864/microsoft-kills-stack-ranking-internal-structure.

137. David A. Garvin, Alison Berkley Wagonfeld, and Liz Kind, "Google's Project Oxygen: Do Managers Matter?" Harvard Business School Case 313-110, April 2013 (revised July 2013).

138. We wouldn't see anything like this again until 2008, when Bill Coughran, our SVP of Research and Systems Infrastructure until 2011, amassed 180 direct reports.

139. Atul Gawande, "The Checklist," *The New Yorker,* December 10, 2007, http://www.newyorker.com/reporting/2007/12/10/071210fa_fact_gawande.

140. From internal interviews.

141. ASTD Staff, "$156 Billion Spent on Training and Development," *ASTD* (blog), American Society for Training and Development (now the Association for Talent Development), December 6, 2012, http://www.astd.org/Publications /Blogs/ASTD-Blog/2012/12/156-Billion-Spent-on-Training-and-Development.

142. "Fast Facts," National Center for Education Statistics, http://nces.ed .gov/fastfacts/display.asp?id=66.

143. Damon Dunn, story told at the celebration of the naming of the William V. Campbell Trophy, Stanford University, Palo Alto, September 8, 2009; http://en.wikipedia.org/wiki/Damon_Dunn.

144. K. Anders Ericsson, "Deliberate Practice and the Acquisition and Maintenance of Expert Performance in Medicine and Related Domains," *Academic Medicine* 79, no. 10 (2004): S70-S81, http://journals.lww.com/academicmedicine /Fulltext/2004/10001/Deliberate_Practice_and_the_Acquisition_and.22.aspx/.

145. Angela Lee Duckworth, Teri A. Kirby, Eli Tsukayama, Heather Berstein, and K. Anders Ericsson, "Deliberate Practice Spells Success: Why Grittier Competitors Triumph at the National Spelling Bee," *Social Psychological and Personality Science* 2, no. 2 (2011): 174–181, http://spp.sagepub.com/content /2/2/174.short.

146. Andrew S. Grove, *High Output Management* (New York: Random House, 1983), 223.

147. Chade-Meng Tan, *Meng's Little Space* (blog), http://chademeng.com/.

148. Jon Kabat-Zinn, *Wherever You Go, There You Are: Mindfulness Meditation in Everyday Life* (New York: Hyperion, 1994), 4.

149. Lucy Kellaway, "The Wise Fool of Google," *Financial Times,* June 7, 2012, http://www.ft.com/intl/cms/s/0/e5ca761c-af34-11e1-a4e0-00144feabdco.html #axzz2dmOsqhuM.

150. Personal conversation.

151. "Teaching Awareness at Google: Breathe Easy and Come into Focus," *Google* (official blog), June 4, 2013, http://googleblog.blogspot.com/search/label/g2g.

152. Michael M. Lombardo and Robert W. Eichinger, *The Career Architect Development Planner* (Minneapolis: Lominger, 1996), iv. Allen Tough, *The Adult's Learning Projects: A Fresh Approach to Theory and Practice in Adult Learning* (Toronto: OISE, 1979).

153. "Social & Environmental Responsibility Report 2011–2012," Gap Inc., http://www.gapinc.com/content/csr/html/employees/career-development.html.

154. "U.S. Corporate Responsibility Report 2013," PricewaterhouseCoopers, http://www.pwc.com/us/en/about-us/corporate-responsibility/corporate-respon sibility-report-2011/people/learning-and-development.jhtml.

155. "Learning at Dell," Dell Inc., http://www.dell.com/learn/au/en/aucorp1 /learning-at-dell.

156. D. Scott DeRue and Christopher G. Myers, "Leadership Development: A Review and Agenda for Future Research," in *The Oxford Handbook of Leadership and Organizations*, ed. David V. Day (New York: Oxford University Press, 2014), http://www-personal.umich.edu/~cgmyers/deruemyersoxfordhand bookcha.pdf.

157. "Kirkpatrick Hierarchy for Assessment of Research Papers," Division of Education, American College of Surgeons, http://www.facs.org/education/tech nicalskills/kirkpatrick/kirkpatrick.html.

158. Yevgeniy Dodis, "Some of My Favorite Sayings," Department of Computer Science, New York University, cs.nyu.edu/~dodis/quotes.html.

159. David Streitfeld, "Silicon Valley's Favorite Stories," *Bits* (blog), *New York Times*, February 5, 2013, http://bits.blogs.nytimes.com/2013/02/05/silicon -valleys-favorite-stories/?_r=0.

160. "William Shockley Founds Shockley Semiconductor," Fairchild Semiconductor Corporation, http://www.fairchildsemi.com/about-fairchild/history/#.

161. Tom Wolfe, "The Tinkerings of Robert Noyce: How the Sun Rose on the Silicon Valley," *Esquire*, December 1983.

162. Nick Bilton, "Why San Francisco Is Not New York," *Bits* (blog), *New York Times*, March 20, 2014, http://bits.blogs.nytimes.com/2014/03/20/why-san -francisco-isnt-the-new-new-york/.

163. Google "Marge vs. the Monorail" for why our Sydney conference room is named North Haverbrook.

164. All images from the Internet Archive, http://archive.org/web/web.php.

165. Comments made during March 2012 interview with *Bloomberg Businessweek* editor Josh Tyrangiel at the 92nd Street Y in Manhattan. See Bianca Bosker, "Google Design: Why Google.com Homepage Looks So Simple," *Huffington Post*, March 27, 2012, http://www.huffingtonpost.com/2012/03/27/google-design -sergey-brin_n_1384074.html.

166. Bosker, "Google Design."

167. Silicon Valley Index, http://www.siliconvalleyindex.org/index.php/econ omy/income.

168. Wayne F. Cascio, "The High Cost of Low Wages," *Harvard Business Review*, December 2006, http://hbr.org/2006/12/the-high-cost-of-low-wages/ar/1.

169. Edward P. Lazear, "Why Is There Mandatory Retirement?" *Journal of Political Economy* 87, no. 6 (1979): 1261–1284.

170. Frank L. Schmidt, John E. Hunter, Robert C. McKenzie, and Tressie W. Muldrow, "Impact of Valid Selection Procedures on Work-Force Productivity," *Journal of Applied Psychology* 64, no. 6 (1979): 609–626.

171. Ernest O'Boyle Jr. and Herman Aguinis, "The Best and the Rest: Revisiting the Norm of Normality of Individual Performance," *Personnel Psychology* 65, no. 1 (2012): 79–119.

172. Nassim Nicholas Taleb, *The Black Swan* (New York: Random House, 2007).

173. Storyboard, "Walt Disney's Oscars," The Walt Disney Family Museum, February 22, 2013, http://www.waltdisney.org/storyboard/walt-disneys-oscars %C2%AE.

174. *Wikipedia*, "List of Best-Selling Fiction Authors," last modified April 19, 2014, http://en.wikipedia.org/wiki/List_of_best-selling_fiction_authors.

175. Correspondence with the Recording Academy.

176. Bill Russell page, *NBA Encyclopedia: Playoff Edition*, National Basketball Association, http://www.nba.com/history/players/russell_bio.html.

177. http://www.golf.com/tour-and-news/tiger-woods-vs-jack-nicklaus-major -championship-records.

178. "Billie Jean King," International Tennis Hall of Fame and Museum, http://www.tennisfame.com/hall-of-famers/billie-jean-king.

179. "Inflation Calculator," *Davemanuel.com*, http://www.davemanuel.com /inflation-calculator.php.

180. I worked with a sales leader who liked to argue that he should get commission based on all of the company's revenue, which was in the billions of dollars. Yes, he was a fantastic salesman, but he would have sold a lot less if he didn't have the company's brand to open doors and give him credibility, its AAA credit rating to provide low costs, and the support of the company's infrastructure. His revenue performance wasn't entirely attributable just to his unique efforts. When implementing "extreme pay," it's essential to discern how much of the exceptional performance is due to the individual and how much is from other factors.

181. Katie Hafner, "New Incentive for Google Employees: Awards Worth Millions," *New York Times*, February 1, 2005, http://www.nytimes.com/2005 /02/01/technology/01google.html?_r=0, http://investor.google.com/corporate/2004 /founders-letter.html.

182. "2004 Founders' Letter," Google: Investor Relations, December 31, 2004, http://investor.google.com/corporate/2004/founders-letter.html.

183. "2005 Founders' Letter," Google: Investor Relations, December 31, 2005, http://investor.google.com/corporate/2005/founders-letter.html.

184. "The Hollywood Money Machine," Fun Industries Inc., http://www .funindustries.com/hollywood-money-blower.htm.

185. John W. Thibaut and Laurens Walker, *Procedural Justice: A Psychological Analysis* (Mahwah, NJ: Lawrence Erlbaum Associates, 1975), http://books.google

.com/books?id=2l5_QgAACAAJ&dq=thibaut+and+walker+1975+Procedural+jus
tice:+A+psychological+analysis.

186. Scott A. Jeffrey, "The Benefits of Tangible Non-Monetary Incentives" (unpublished manuscript, University of Chicago Graduate School of Business, 2003), http://theirf.org/direct/user/site/o/files/the%20benefits%20of%20tangible %20non%20monetary%20incentives.pdf. Scott A. Jeffrey and Victoria Shaffer, "The Motivational Properties of Tangible Incentives," *Compensation & Benefits Review* 39, no. 3 (2007): 44–50. Erica Mina Okada, "Justification Effects on Consumer Choice of Hedonic and Utilitarian Goods," *Journal of Marketing Research* 42, no. 1 (2005): 43–53. Richard H. Thaler, "Mental Accounting Matters," *Journal of Behavioral Decision Making* 12, no. 3 (1999): 183–206.

187. This finding is consistent with the academic work, which focuses on purchases rather than gifts. People are happier when they buy experiences (trips, dinners) than when they buy things (clothes, electronics). Travis J. Carter and Thomas Gilovich, "The Relative Relativity of Material and Experiential Purchases," *Journal of Personality and Social Psychology* 98, no. 1 (2010): 146–159.

188. Adam Bryant, "Honeywell's David Cote, on Decisiveness as a 2-Edged Sword," *New York Times*, November 2, 2013, http://www.nytimes.com/2013/11/03 /business/honeywells-david-cote-on-decisiveness-as-a-2-edged-sword.html.

189. Ben Parr, "Google Wave: A Complete Guide," *Mashable*, May 28, 2009, last updated January 29, 2010, http://mashable.com/2009/05/28/google -wave-guide/.

190. "Introducing Apache Wave," Google, *Google Wave Developer Blog*, December 6, 2010, http://googlewavedev.blogspot.com/2010/12/introducing-apache -wave.html.

191. Chris Argyris, "Double Loop Learning in Organizations," *Harvard Business Review*, September 1977, http://hbr.org/1977/09/double-loop-learning -in-organizations/ar/1.

192. Chris Argyris, "Teaching Smart People How to Learn," *Harvard Business Review*, May 1991, http://hbr.org/1991/05/teaching-smart-people-how-to-learn/.

193. This echoes a line purportedly uttered by IBM founder Thomas J. Watson Sr.: "Recently, I was asked if I was going to fire an employee who made a mistake that cost the company $600,000. No, I replied, I just spent $600,000 training him. Why would I want somebody to hire his experience?"

194. "California Middle School Rankings," *SchoolDigger.com*, http://www .schooldigger.com/go/CA/schoolrank.aspx?level=2. SchoolDigger's methodology ranks schools based on the sum of the average math and English test scores on state standardized tests.

195. Dave Eggers, *The Circle* (New York: Knopf, 2013).

196. Ronald S. Burt, "Structural Holes and Good Ideas," *American Journal of Sociology* 110, no. 2 (2004): 349–399.

197. I'll take whoever of the two is available. . . . I don't need two people to cut my hair.

198. Nicholas Carlson, "Marissa Mayer Sent a Late Night Email Promising to Make Yahoo 'the Absolute Best Place to Work' (YHOO)," *SFGate*, August 27, 2012, http://www.sfgate.com/technology/businessinsider/article/Marissa-Mayer -Sent-A-Late-Night-Email-Promising-3817913.php.

199. Jillian Berman, "Bring Your Parents to Work Day Is a Thing. We Were There," *Huffington Post*, November 11, 2013, http://www.huffingtonpost.com /2013/11/11/take-parents-to-work_n_4235803.html.

200. Meghan Casserly, "Here's What Happens to Google Employees When They Die," *Forbes*, August 8, 2012, http://www.forbes.com/sites/meghancasserly /2012/08/08/heres-what-happens-to-google-employees-when-they-die/. Meghan's understanding of this and other issues was so thoughtful and impressive that, at the first chance we had, I encouraged our team to recruit her to join Google, which she did.

201. Private conversation.

202. Kahneman won the Nobel Prize for work done with Tversky, but Tversky passed away before it was awarded. Unfortunately, the Nobel is not awarded posthumously. At his Nobel acceptance speech, Kahneman's first words were, "The work on which the award was given...was done jointly with Amos Tversky during a long period of unusually close collaboration. He should have been here." Prize Lecture by Daniel Kahneman, Stockholm University, December 8, 2002, http://www.nobelprize.org/mediaplayer/?id=531.

203. "Inflation Calculator."

204. Amos Tversky and Daniel Kahneman, "The Framing of Decisions and the Psychology of Choice," *Science* 211, no. 4481 (January 30, 1981): 453–458, http://psych.hanover.edu/classes/cognition/papers/tversky81.pdf.

205. Stephen Macknik and Susana Martinez-Conde, *Sleights of Mind: What the Neuroscience of Magic Reveals About Our Everyday Deceptions* (New York: Henry Holt, 2010), 76–77.

206. Julie L. Belcove, "Steamy Wait Before a Walk in a Museum's Rain," *New York Times*, July 17, 2013, http://www.nytimes.com/2013/07/18/arts/steamy-wait -before-a-walk-in-a-museums-rain.html.

207. Michael Barbaro, "The Bullpen Bloomberg Built: Candidates Debate Its Future," *New York Times*, March 22, 2013, http://www.nytimes.com/2013/03/23 /nyregion/bloombergs-bullpen-candidates-debate-its-future.html.

208. Chris Smith, "Open City," *New York*, September 26, 2010, http://nymag .com/news/features/establishments/68511/.

209. Richard H. Thaler and Cass R. Sunstein, *Nudge* (New Haven, CT: Yale University Press, 2008), 15.

210. An obvious difference between a nudge and a bonus plan is that the former is often not disclosed, while a bonus plan is explicitly set up to drive certain behaviors. But once you concede that a company can legitimately shape its employees' behaviors, then you're left with a more difficult question of where exactly the company crosses the line from "good" shaping to "bad" shaping. I'd argue that that line is influenced by how transparent the company is about its nudges.

211. They were. George Musser, an editor at *Scientific American*, wrote in the August 17, 2009, issue that cubicles were a reaction to the open floor-plan offices common before the 1950s. Cubicles were meant to give individuals more privacy. George Musser, "The Origin of Cubicles and the Open-Plan Office," *Scientific American*, August 17, 2009, http://www.scientificamerican.com/article.cfm?id=the-origin-of-cubicles-an/.

212. Bradley Johnson, "Big U.S. Advertisers Boost 2012 Spending by Slim 2.8% with a Lift from Tech," *Advertising Age*, June 23, 2013, http://adage.com/article/news/big-u-s-advertisers-boost-2012-spending-slim-2-8/242761/.

213. Special Issue: U.S. Beverage Results for 2012, *Beverage Digest*, March 25, 2013, http://www.beverage-digest.com/pdf/top-10_2013.pdf.

214. Samuel M. McClure, Jian Li, Damon Tomlin, Kim S. Cypert, Latané M. Montague, and P. Read Montague, "Neural Correlates of Behavioral Preference for Culturally Familiar Drinks," *Neuron* 44, no. 2 (2004): 379–387.

215. Nyla R. Branscombe, Naomi Ellemers, Russell Spears, and Bertjan Doosje, "The Context and Content of Social Identity Threat," in *Social Identity: Context, Commitment, Content*, eds. Naomi Ellemers, Russell Spears, and Bertjan Doosje (Oxford, UK: Wiley-Blackwell, 1999), 35–58.

216. Robert B. Cialdini, "Harnessing the Science of Persuasion," *Harvard Business Review* 79, no. 9 (2001): 72–81, http://lookstein.org/leadership/case-study/harnessing.pdf.

217. Bradford D. Smart, *Topgrading: How Leading Companies Win by Hiring, Coaching, and Keeping the Best People* (Upper Saddle River, NJ: Prentice Hall, 1999).

218. Autumn D. Krauss, "Onboarding the Hourly Workforce." Poster presented at the Society for Industrial and Organizational Psychology (SIOP), Atlanta, GA, 2010.

219. "Surgical Safety Checklist (First Edition)," World Health Organization, http://www.who.int/patientsafety/safesurgery/tools_resources/SSSL_Checklist_finalJun08.pdf.

220. Alex B. Haynes et al., "A Surgical Safety Checklist to Reduce Morbidity and Mortality in a Global Population," *New England Journal of Medicine* 360 (2009): 491–499, http://www.nejm.org/doi/full/10.1056/NEJMsa0810119.

221. Michael Lewis, "Obama's Way," *Vanity Fair*, October 2012, http://www.vanityfair.com/politics/2012/10/michael-lewis-profile-barack-obama.

222. Talya N. Bauer, "Onboarding New Employees: Maximizing Success," SHRM Foundation's Effective Practice Guidelines (Alexandria, VA: SHRM Foundation, 2010), https://docs.google.com/a/pdx.edu/file/d/0B-bOAWJkyKwUMzg2YjE3MjctZjko0C0oZmFiLWFiMmMtYjFiMDdkZGE4MTY3/edit?hl=en_US&pli=1.

223. Susan J. Ashford and J. Stewart Black, "Proactivity During Organizational Entry: The Role of Desire for Control," *Journal of Applied Psychology* 81, no. 2 (1996): 199–214.

224. There's a sizable body of evidence showing that proactive employees are better performers across industries. B. Fuller Jr. and L. E. Marler, "Change Driven by Nature: A Meta-Analytic Review of the Proactive Personality," *Journal*

of Vocational Behavior 75, no. 3 (2009): 329–345. (A meta-analysis of 107 studies.) Jeffrey P. Thomas, Daniel S. Whitman, and Chockalingam Viswesvaran, "Employee Proactivity in Organizations: A Comparative Meta-Analysis of Emergent Proactive Constructs," *Journal of Occupational and Organizational Psychology* 83, no. 2 (2010): 275–300. (A meta-analysis of 103 samples.)

225. *Wikipedia*, "Poka-yoke," last modified May 11, 2014, http://en.wikipedia.org/wiki/Poka-yoke.

226. Steven F. Venti and David A. Wise, "Choice, Chance, and Wealth Dispersion at Retirement," in *Aging Issues in the United States and Japan*, eds. Seiritsu Ogura, Toshiaki Tachibanaki, and David A. Wise (Chicago: University of Chicago Press, 2001), 25–64.

227. *Wikipedia*, "Household Income in the United States," http://en.wikipedia.org/wiki/Household_income_in_the_United_States. Carmen DeNavas-Walt, Bernadette D. Proctor, and Jessica C. Smith, "Income, Poverty, and Health Insurance Coverage in the United States: 2011," US Census Bureau (Washington, DC: US Government Printing Office, 2012). "Supplemental Nutrition Assistance Program (SNAP)," United States Department of Agriculture, http://www.fns.usda.gov/pd/snapsummary.htm. J. N. Kish, "U.S. Population 1776 to Present," https://www.google.com/fusiontables/DataSource?dsrcid=225439.

228. Chart from Venti and Wise, "Choice, Chance, and Wealth."

229. Ibid., 25.

230. B. Douglas Bernheim, Jonathan Skinner, and Steven Weinberg, "What Accounts for the Variation in Retirement Wealth among U.S. Households?" *American Economic Review* 91, no. 4 (2001): 832–857, http://www.econ.wisc.edu/~scholz/Teaching_742/Bernheim_Skinner_Weinberg.pdf.

231. James J. Choi, Emily Haisley, Jennifer Kurkoski, and Cade Massey, "Small Cues Change Savings Choices," National Bureau of Economic Research Working Paper 17843, revised June 29, 2012, http://www.nber.org/papers/w17843.

232. Richard H. Thaler and Shlomo Benartzi, "Save More Tomorrow: Using Behavioral Economics to Increase Employee Savings," *Journal of Political Economy* 112, no. 1 (2004): S164–S187, http://faculty.chicagobooth.edu/Richard.Thaler/research/pdf/SMarTJPE.pdf.

233. Yes, it's really trademarked.

234. Todd had no way of knowing that later that year we'd announce Calico, a Google business led by Art Levinson, the former CEO of Genentech, with a goal of addressing the debilitating and inevitable consequences of aging.

235. "Obesity and Overweight," National Center for Health Statistics, Centers for Disease Control and Prevention, last updated May 14, 2014, http://www.cdc.gov/nchs/fastats/overwt.htm.

236. "Overweight and Obesity: Adult Obesity Facts," Centers for Disease Control and Prevention, last updated March 28, 2014, http://www.cdc.gov/obesity/data/adult.html.

237. M. Muraven and R. F. Baumeister, "Self-Regulation and Depletion of Limited Resources: Does Self-Control Resemble a Muscle?" *Psychological Bulletin* 126, no. 2 (2000): 247–259.

238. D. Hammond, G. T. Fong, P. W. McDonald, K. S. Brown, and R. Cameron, "Graphic Canadian Cigarette Warning Labels and Adverse Outcomes: Evidence from Canadian Smokers," *American Journal of Public Health* 94, no. 8 (2004): 1442–1445.

239. Julie S. Downs, Jessica Wisdom, Brian Wansink, and George Loewenstein, "Supplementing Menu Labeling with Calorie Recommendations to Test for Facilitation Effects," *American Journal of Public Health* 103, no. 9 (2013): 1604–1609.

240. "McDonald's USA Nutrition Facts for Popular Menu Items," McDonalds .com, effective May 27, 2014, http://nutrition.mcdonalds.com/getnutrition/nutri tionfacts.pdf.

241. David Laibson, "A Cue-Theory of Consumption," *Quarterly Journal of Economics* 116, no. 1 (2001): 81–119.

242. Colleen Giblin, "The Perils of Large Plates: Waist, Waste, and Wallet," review of "The Visual Illusions of Food: Why Plates, Bowls, and Spoons Can Bias Consumption Volume," by Brian Wansink and Koert van Ittersum (*FASEB Journal* 20, no. 4 [2006]: A618), Cornell University Food and Brand Lab, 2011, http:// foodpsychology.cornell.edu/outreach/large-plates.html.

243. Wansink and Ittersum, "Visual Illusions of Food."

244. Leo Benedictus, "The Nudge Unit—Has It Worked So Far?" *Guardian*, May 1, 2013, http://www.theguardian.com/politics/2013/may/02/nudge-unit -has-it-worked.

245. Britton Brewer, "Adherence to Sport Injury Rehabilitation Regimens," in *Adherence Issues in Sport and Exercise*, ed. Stephen Bull (New York: Wiley, 1999), 145–168.

246. Richard H. Thaler, "Opting In vs. Opting Out," *New York Times*, September 26, 2009, http://www.nytimes.com/2009/09/27/business/economy/27view.html.

247. Eric J. Johnson and Daniel Goldstein, "Do Defaults Save Lives?," *Science* 302, no. 5649 (2003): 1338–1339.

248. Zechariah Chafee Jr., "Freedom of Speech in War Time," *Harvard Law Review* 32, no. 8 (1919): 932–973, http://www.jstor.org/stable/1327107?seq=26&.

249. "Our Work: What We Believe," McKinsey & Company, http://www .mckinsey.com.br/our_work_belive.asp.

250. Andrew Hill, "Inside McKinsey," *FT Magazine*, November 25, 2011, http:// www.ft.com/cms/s/2/0d506e0e-1583-11e1-b9b8-00144feabdc0.html#axzz2iCZ5ks73.

251. Ralph Waldo Emerson, "Self-Reliance," *Essays* (1841), republished as *Essays: First Series* (Boston: James Munroe and Co., 1847).

252. http://googleblog.blogspot.com/2011/07/more-wood-behind-fewer -arrows.html.

253. Mixed metaphor, I know. I use it because I find that few management practices are completely binary. For example, few companies say "innovate always in everything we do" or "never innovate." Instead, management practices gather strength over time before ossifying and becoming dysfunctional. Companies organize by geography and then realize their products don't work the same in every region and are too expensive to support, so then they reorganize along product lines. Then the products stop being well-suited to local needs so they

reorganize again. Executive management is the art of knowing when it's time to swing the pendulum back the other way.

254. *Wikipedia*, "Goji," http://en.wikipedia.org/wiki/Goji.

255. Jonathan Edwards, "Sinners in the Hands of an Angry God. A Sermon Preached at Enfield, July 8th, 1741," ed. Reiner Smolinski, Electronic Texts in American Studies Paper 54, Libraries at University of Nebraska–Lincoln, http://digitalcommons.unl.edu/cgi/viewcontent.cgi?article=1053&context=etas.

256. Steven Pinker, "Violence Vanquished," *Wall Street Journal*, September 24, 2011, http://online.wsj.com/news/articles/SB10001424053111190410670457658 3203589408180.

257. United States Congress House Special Committee to Investigate the Taylor and Other Systems of Shop Management, *The Taylor and Other Systems of Shop Management: Hearings before Special Committee of the House of Representatives to Investigate the Taylor and Other Systems of Shop Management* (Washington, DC: US Government Printing Office, 1912), 3: 1397, http://books.google.com/books?id=eyrbA AAAMAAJ&pg=PA1397&lpg=PA1397&dq=physically+able+to+handle+pig-iron.

258. Tony Hsieh: "I think of myself less as a leader, and more of being almost an architect of an environment that enables employees to come up with their own ideas, and where employees can grow the culture and evolve it over time." (Adam Bryant, "On a Scale of 1 to 10, How Weird Are You?," *New York Times*, January 9, 2010.)

Reed Hastings: "Responsible people thrive on freedom and are worthy of freedom. Our model is to increase employee freedom as we grow, rather than limit it, to continue to attract and nourish innovative people, so we have a better chance of sustained success." ("Netflix Culture: Freedom and Responsibility," August 1, 2009, http://www.slideshare.net/reed2001/culture-1798664.)

During the recession that started in 2008, Jim Goodnight asked SAS employees to find their own ways to steer the company through the downturn: "I told them we would have no layoffs for the entire year—but that I needed them to pitch in and reduce expenses, to slow down on hiring and cut it out completely if possible. Everybody did pitch in and productivity actually went up in 2009....It was one of our top three most profitable years." ("SAS Institute CEO Jim Goodnight on Building Strong Companies—and a More Competitive U.S. Workforce," *Knowledge@Wharton*, Wharton School of the University of Pennsylvania, January 5, 2011, http://bit.ly/1dyJMoJ.)

259. Abraham H. Maslow, "A Theory of Human Motivation," *Psychological Review* 50, no. 4 (July 1943): 370–396. Maslow's hierarchy, though well known, ultimately failed to be supported by data. Others have worked to refine Maslow's work, including Douglas T. Kenrick, Vladas Griskevicius, Steven L. Neuberg, and Mark Schaller, who offered an updated framework in 2010 ("Renovating the Pyramid of Needs," *Perspectives on Psychological Science* 5, no. 3 [2010]: 292–314, http://pps.sagepub.com/content/5/3/292.short).

Index

About the Author

Laszlo Bock is Senior Vice President of People Operations for Google, Inc. He leads Google's people function, which includes all areas related to the attraction, development, and retention of "Googlers," of which there are more than fifty thousand in more than seventy offices worldwide. During Laszlo's tenure Google has been recognized over a hundred times as an exceptional employer, including being named the #1 Best Company to Work For across the United States, Argentina, Australia, Brazil, Canada, France, India, Ireland, Italy, Japan, Korea, Mexico, the Netherlands, Poland, Russia, Switzerland, and the United Kingdom; the #1 Top Diversity Employer; the most desirable employer for undergraduates, MBAs, and college graduates in numerous countries; the best company for women in technology; and honors such as a perfect score from the Human Rights Campaign.

Laszlo also worked at the General Electric Company and McKinsey & Company, serving clients in the technology, private equity, and media industries on a wide range of strategic and operational issues, including growth and turnaround strategy. Earlier, he had worked at another consulting firm, a start-up, as an actor, and cofounded a nonprofit organization working with at-risk youth. He is a member of the board of trustees of Pomona College and has served as an advisor or board member of several venture capital-funded companies.

He earned a bachelor's degree in international relations from Pomona College and an MBA from the Yale School of Management.

Laszlo has testified before Congress on immigration reform

and labor issues. He has been featured in the *Wall Street Journal*, the *New York Times*, the *Washington Post*, and on PBS *NewsHour* and the *Today* show.

In 2010 he was named "Human Resources Executive of the Year" by *Human Resource Executive* magazine. In 2014 they named him one of the "ten most influential people impacting HR" for the decade, the only HR executive to be named to the list. He (briefly) co-held the world record for Greek Syrtaki dance along with 1,671 other Googlers. On November 3, 2012, he was ranked number one in the world in daily total kills in Assassin's Creed III on Xbox. He owns a lot of comic books. A lot.